ACKNOWLEDGMENTS

This book represents many months of work and there are a great number of people I would like to thank for their patience and encouragement. To Ellen, whom I love and admire for listening to my complaints, being my comfort, keeping me in pants,and for being my keeper, thank you. Vonja and Deborah Kirkland, who deserve a standing ovation for their unwavering moral and financial support; God sent you to me. I would like to thank other family members, including friends who have only had good things to say in the midst of all my anxiety: Wilson McCord Sr. and Peggy, Tsa and Austin Ross McCord, Catherine Amendolara and Don Campbell, Hans V., Brenda H., Steven S., Craig M., John C., Craig S., Angela L., Willow for company on those cold winter nights, and Kitty, Julie, and Pumpkin for listening and maybe not quite understanding. Ron David and Glenn Thompson, thanks for trusting me with this project; I have learned a great deal. My very special thanks go to Patricia Allen, for her help, in making the manuscript like butter.

021430

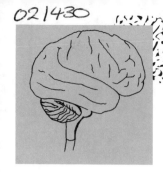

TABLE OF
CONTENTS

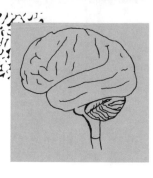

ISLE COLLEGE
RESOURCES CENTRE

BIOLOGY
FOR BEGINNERS™

BY WILBUR McCORD JR.

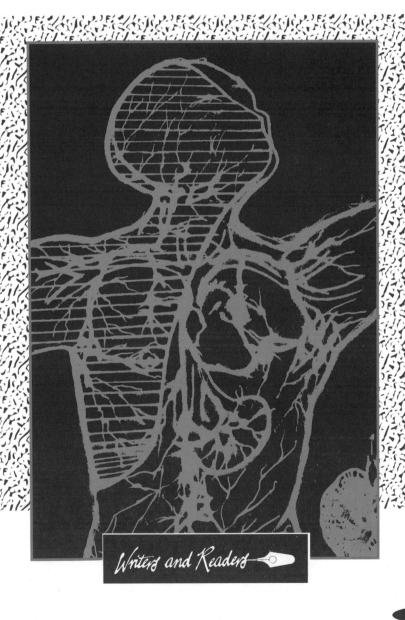

Writers and Readers

WRITERS AND READERS PUBLISHING, INC.

P.O. Box 461, Village Station
New York, NY 10014

Writers and Readers Limited
9 Cynthia Street
London N1 9JF
England

•

A Writers and Readers Documentary Comic Book
Copyright © 1995
ISBN # 0-86316-194-4
1 2 3 4 5 6 7 8 9 0

Manufactured in the United States of America

Beginners Documentary Comic Books are published by Writers and Readers Publishing, Inc. Its trademark, consisting of the words "For Beginners, Writers and Readers Documentary Comic Books" and the Writers and Readers logo, is registered in the U. S. Patent and Trademark Office and in other countries.

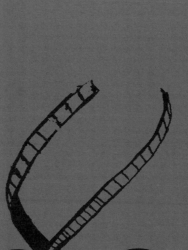

BIOLOGY

FOR BEGINNERS™

This book is
dedicated to all
people living with
the HIV virus.

A Brief History of Life: From Our Very Beginnings to Modern Times

Biology is the study of life. And humans have been trying to understand what life really is for thousands of years. The methods and the information we have accumulated about life began with our ancestors in prehistory.

Artifacts, or things left behind, tell us something about our ancestors' observations of the living world. Cave paintings, like those at Lascaux, France, give us an idea about our ancestors' world. They also give us insights into their understanding of that world.

The paintings are said to be between 12,000 and 17,000 years old. Interpretations of the scenes at Lascaux vary. A number of archaeologists think that some of the scenes were painted for religious reasons. But it really does not matter, for our purposes, whether or not the prehistoric renderings were documented for religious or secular purposes. We'll never know if they are records of events or stories that were told for each passing generation.

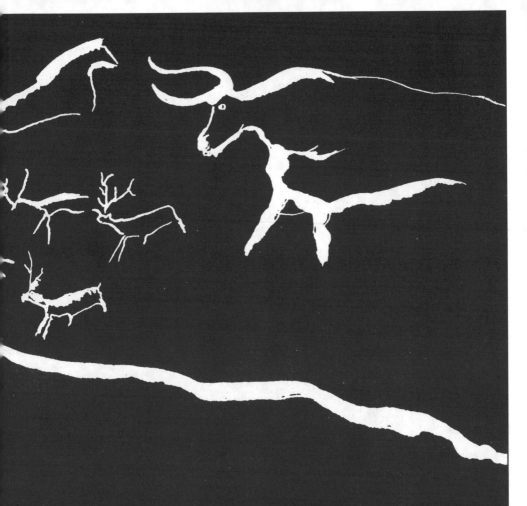

 It might have even been a lesson in
hunting. What we can say, after
looking at these paintings is, that
we have always been aware of our
natural surroundings. We have
always lived with and always will be a
part of nature. All living animals,
humans included, are faced with the
day to day task of survival. From
that task of survival humans first
learned the ways of the natural
world.

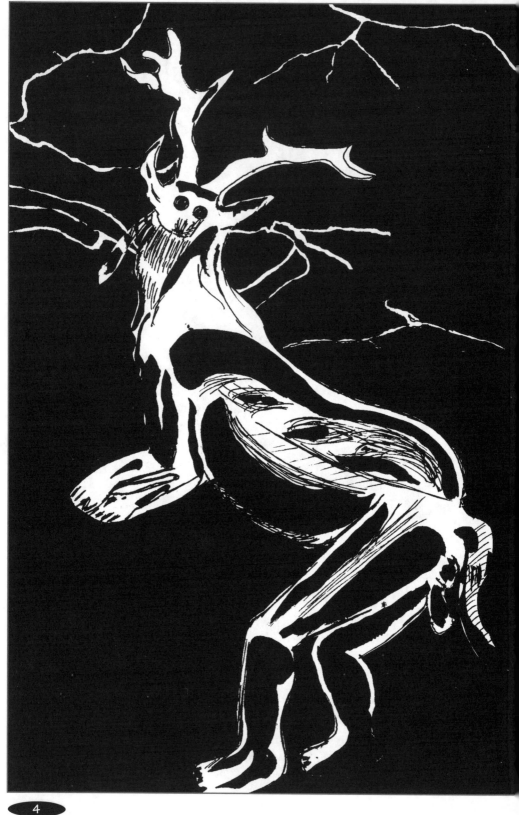

We used magic to help us convince the forces of nature to bring rain, so that our crops would grow. We also called on the unknown to help cure human illness. And in this way medicine, the alteration of the human organism to help cure disease, began.

As early as 10,000 years ago, we recognized that rain had an affect on the plant foods. This recognition of the link between water and growth is one example of the relationships in nature that began our search for an understanding of the living world.

Magic allowed us to give meaning to natural phenomena. Spirits that inhabited and controlled the living world could even be coaxed. Disaster was explained in their anger. Medicinal plants were tried on many humans. The record of their successes and failures was kept by the medicine men or women and passed on. Many of our explorations of plant foods and medicines came from watching other animals who taught us what we might be able to ingest. We learned slowly and, today, are still learning to cooperate with nature.

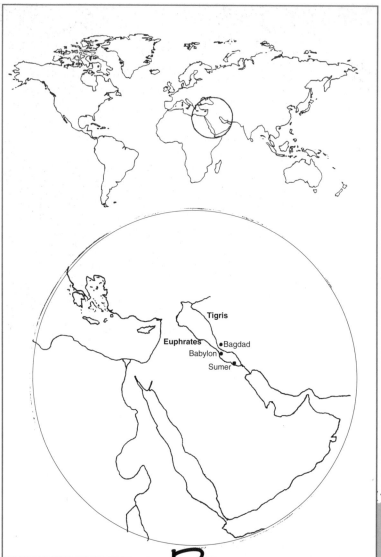

Tigris

Euphrates
Bagdad
Babylon
Sumer

Between the Tigris and Euphrates Rivers, on the foothills of the fertile crescent, the first city-state rose in Southern Mesopotamia. This ancient region of Southwest Asia was the cradle of civilization and is known today as the country of Iraq. Here, the first system of writing was developed, and it is here that the first cities were built.

The picture symbols developed during the Sumerian period (before 3000 B.C.) were written on clay tablets. It was a pictographic writing system with over 2000 signs representing symbols. Hundred of years later, the Egyptians would only use a little over 700 symbols in their written language.

Sumer was the first great civilization of mankind (around 4000 B.C.). During his reign, which began in 2340 B.C., Sargon the Great united the city states in the south. This period is known as the Akkadian Dynasty. The Akkadian language adopted the cuneiform writing symbols of the Sumerians.

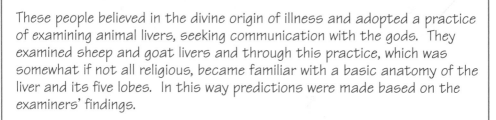

Between the years 1792 to 1750 B.C., the city state of Babylon was ruled by its most famous ruler, Hammurabi. The stele on which his code of laws is written provides a fundamental survey of plants and animals in the area. In their writings, it has been found that the people used plants for drugs, but it is difficult to identify each plant used and the illness for which it was prescribed. Medical information was available, but much depended on the physician's experience. In this code of laws we also find the first regulations for medical practice. The code also mentions the use of instruments for surgery. And archaeologists have recovered surgical tools from the age of Hammurabi.

These people believed in the divine origin of illness and adopted a practice of examining animal livers, seeking communication with the gods. They examined sheep and goat livers and through this practice, which was somewhat if not all religious, became familiar with a basic anatomy of the liver and its five lobes. In this way predictions were made based on the examiners' findings.

The Mesopotamians collected a great deal of information on animals and during Babylonian times attempted a system of classification. Tablets have been found that give the names of hundreds of animals and specific information on plants in the area. Shellfish were distinguished from other water animals and fish, and these were separated from animals that lived on land. Large groups such as four-legged creatures were further broken down. Plants that bore fruit and those that bore nuts, were also separated, as were the flowering plants and the non-flowering plants. These people discovered that the date palm has a female and male plant, and that there would be no crop if the two plants were not side by side. This same phenomenon was discovered in mandrakes and in cypress trees.

The association made between the curing of illness and the symbol of the snake reaches back to Mesopotamia, for Ningishzide, the double-headed snake, is the emblem for medicine , which is even used today. In the epic *Gilgamesh*, it is the snake who steals and eats the plant of eternal life; the snake sheds its skin and is rejuvenated. Thus it became the symbol of the cure for disease. Sickness at that time was seen as a curse, a punishment by the gods. Illness that followed a family generation after generation would not be recognized as genetic for thousands of years.

Some scholars think that it was from Mesopotamia that the Egyptian culture arose in the west. Pre-dynastic tablets recently unearthed resemble the cuneiform tablets of late Mesopotamia.

The Rosetta Stone, carved in hieroglyph-ics and repeated in Greek and Demotic or simple characters, was discovered in 1799 when Napoleon conquered Egypt. Egyptian hieroglyphics were not under-standable until this discovery which allowed scholars to translate the texts. A number of ancient texts found described the vessels, movement, and structure of the heart.

It is known that the embalmers who pre-pared the dead removed the internal organs. They were familiar with the anatomy of the human body, but their understanding of anatomy and physiology were a part of the Egyptian religions.

The Egyptians knew that the beat of the heart could be felt throughout the body. Their observations of the brain have been found as well as writings, where an author describes the brain as the body's control center. Information on the motor func-tions of the brain and defects caused by injury have been found. But, like those before them and many to come, the Egyptians misinterpreted most of what they documented.

As early as the 4th century B.C., the Chinese described the diversity of animals in their region. They selectively bred animals and plants and wrote pharmaceutical outlines. They compiled all of the information in the form of encyclopedias.

Knowledge of silkworm breeding has been cited as far back as 1500 B.C.. The Chinese bred insects for medicinal ingredients and to protect crops. Citrus farmers hung bags of ants on their trees to protect the fruit from insects that destroyed the crop. This was the first use of biological plant protection. Today, we use the same species of ant to protect our own citrus crop.

The study of plants in China continued throughout their history. In the West, the study of plants began with the Greeks and did not continue until the Renaissance. The Chinese detailed different species of plants and trees and studied the soils of each. They even grew edible plants outside of their natural habitats.

Long before Linneaus in the 18th century, the Chinese had attempted a system of plant (botanical) classification. The Chinese used only their language to name plants through the ages. Even then two-word technical names were used to clarify and recognize the families of plants.

A

cupuncture and medical treatment has been used by the Chinese for thousands of years to balance out the yin or yang in humans. They pierce the skin with long needles to transmit the life force chi. With each point on the body corresponding to specific internal organs all illnesses considered correctable by acupuncture. But, Confucius (K'ung Futzu, 550-479 B.C.), who set the standards of behavior, prohibited the study of anatomy and physiology until the 18th century.

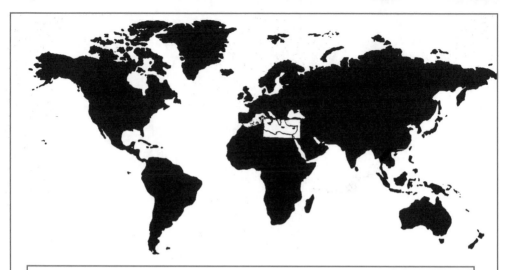

Scientific Philosophy in Greece

Perganon

Chios

Miletus

Cnidos

Alexandria

Hundreds of years before the existence of the Greek Ionian colonies in the 6th century B.C., there had been contact and an exchange of ideas between the West and Egypt, Mesopotamia, and probably China. Information on the natural world, experiences, understandings, and misconceptions were passed to the West from those great civilizations. And it was during the 6th century B.C. in Greece that natural philosophy began. This was a new approach to understanding nature.

The Iron Age (in Europe, c.1600 B.C.) had recently changed Western civilization and increased trade with the Near East. An alphabet had spread to all areas of the Greek cultures by 1000 B.C. And it is in one of these colonies that the first scientist philosopher was born in the West.

His name was Thales, and he was born around 640 B.C. in Miletus on the Aegean west coast of Asia Minor, then the outskirts of the Greek world. It was like other areas that had been growing over the centuries, where knowledge was passed down orally from each succeeding generation.

Thales is known as the "father of science," because he explained phenomena without depending on supernatural explanations. Though he believed in the gods, he never explained humans or the universe in religious terms.

Thales had a great number of interests and influenced many contemporaries and students alike. He made contributions to many of the sciences existing during his time. Thales believed that water was the basic influence of all life on earth, and that from water came earth and air.

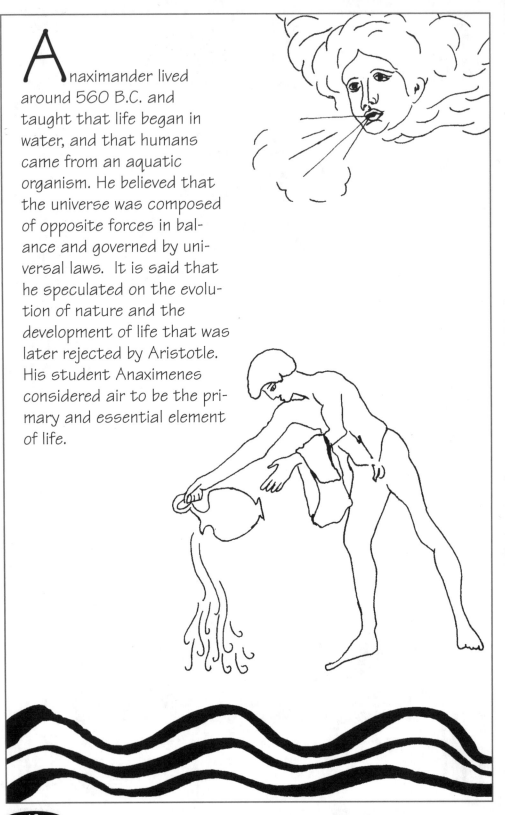

Anaximander lived around 560 B.C. and taught that life began in water, and that humans came from an aquatic organism. He believed that the universe was composed of opposite forces in balance and governed by universal laws. It is said that he speculated on the evolution of nature and the development of life that was later rejected by Aristotle. His student Anaximenes considered air to be the primary and essential element of life.

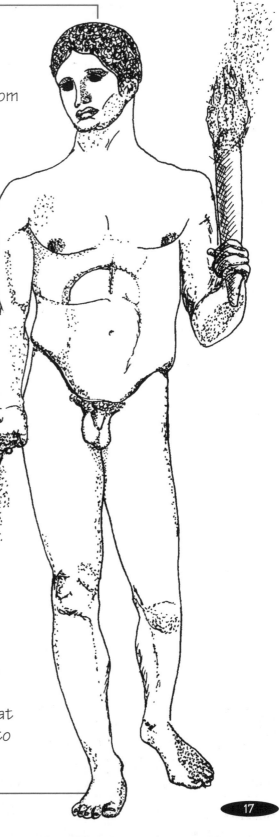

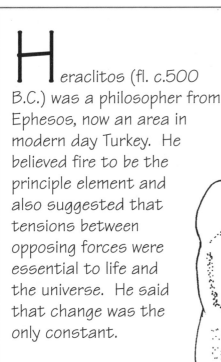 eraclitos (fl. c.500 B.C.) was a philosopher from Ephesos, now an area in modern day Turkey. He believed fire to be the principle element and also suggested that tensions between opposing forces were essential to life and the universe. He said that change was the only constant.

In the 6th century B.C., the four basic elements; air, water, fire, and earth were accepted as components of everything. Later these elements would be converted to the four humors in medical theory.

It was in ancient Greece that the attitude of Western scientific culture grew. Their philosophers adopted a way of looking at nature that had nothing to do with magic.

Hippocrates (born 420 B.C.) lived and taught on the Greek island of Cos, not far from the Turkish coast. The *Corpus Hippocraticum* is a writing attributed to Hippocrates and other medical professionals of his time. Some scholars think that some of these writings actually came from Cnidos, on the Greek mainland. Cnidos had a group of teachers and students that were probably as important as those on Cos. Cnidos was also established earlier than the school on Cos. But the collection of writings on medicine from Cnidos did not survive.

It is really not known if the teachings that came from Hippocrates are of one or many men. These works attributed to him were later collected in the great library of Alexandria in the 4th century B.C. The *Hippocratic Corpus* consisted of 70 works and had a great influence on antiquity and the middle ages. The *Corpus* and physicians in general stressed the importance of the healing power in nature. He taught that the physician aids nature in the body's recovery. He also developed the doctrine of the four humors. In man there was blood, phlegm, yellow bile, and black bile. If proportioned correctly, humans would enjoy perfect health.

The physicians who learned the Hippocratic method of healing were trained so that their experiences became the most important basis of medicine. They were taught to recognize certain diseases and predict accurately its course.

And as his work projects, Hippocrates was said to be an extraordinarily caring man. His life is the example that all doctors strive for, and that all who are sick look for in their doctors. Today, it is the Hippocratic oath that is taken by all doctors at their graduation ceremony.

Alcmaeon of Croton, who lived about the same time as Hippocrates, is known as the first medical scientist. He made direct observations, tested experimentally, and did many dissections, establishing the connection between the sense organs and the brain. He clearly showed the optic nerves and their crossings. Far from being accurate, Alcmaeon felt that humans went to sleep when the blood vessels in the brain drained, and that we awoke when they were filled again. In his report, he also stated that the eye contained fire and water. He condensed the already accepted belief that the senses originated in the brain. Later, Aristotle, one of the greatest thinkers of the ancient Greek era would disagree with him and state that the heart was the center of sensation.

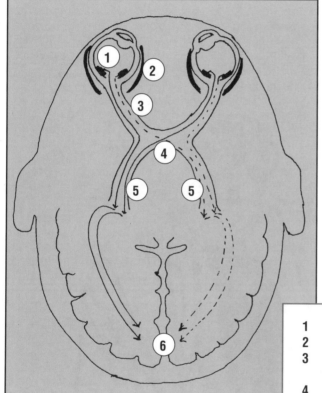

1 eyeball
2 rectus muscles
3 optic nerves (contain optic fiber bundles
4 optic chiasma (neural fibers carrying optic messages cross here)
5 allows messages from each eye to reach the visual cortex in the left and right side of the brain
6 brain

ARISTOTLE

During the 4th century in Greek Antiquity, natural history, particularly zoology, reached its highest point with the philosopher Aristotle (384-322 B.C.). He was a product of Plato's Academy. Later one of his students would initiate the science of botany (the study of plants). It is the first time these subjects would be studied apart from the encompassing natural philosophy.

Aristotle studied in Athens with his father, a physician, when he was 17 years old. He studied and stayed with Plato until his teacher's death and then traveled to Asia Minor to study. Later, he worked as a tutor to the young Alexander in the court of Phillip in Macedonia. He founded the Lyceum on his return to Athens in 335 B.C. and it became the first great scientific research center in history. Before this, Plato's Academy had been a place of philosophical inquiry. The Lyceum was a school for collecting and organizing scientific data.

One of Aristotle's students wrote that Aristotle's progression was from a Platonic and metaphysical one to an empirical and research-based inquiry. His research lead him to organize zoological collections and information. He began to organize the collections with his student, Theophrastos, who later became his associate and successor. As he organized the collections, he distinguished cetaceans (dolphins and whales) from fish. He noted that cetaceans have lungs or blowholes, and that most of the embryos were tied to the mother by one umbilical cord. He also observed the live birth of these animals and knew they were different from fish.

Aristotle wrote books on the anatomy of animals and their life histories. His natural history of animals included the observations and description of individual animal life history. He wrote on their motion and changed certain myths because of his studies. He stressed the importance of collecting a number of facts in order to make conclusions rather than using reason. Aristotle knew that there sometimes were not enough facts to make a conclusion, and that more information could be found by dissecting.

There is evidence that Aristotle not only dissected dead animals but also dissected and described the movement of a tortoise heart. The bee habitat, the development of the chick embryo, the change of the caterpillar to a butterfly and the testes of animals all came under his inquiry. In his observations of embryos he adopted the doctrine of *epigenesis*, which states that the embryo develops from undifferentiated material.

He concerned himself with the sameness and differences of animals and made the most general division possible between red-blooded and non-red-blooded animals. He made the distinction between the vertebrates (animals with backbones) and invertebrates (animals without backbones). His terms *eidos* (species) and *genos* (genus) are the basis of biological classification today.

With closely allied groups differentiated by graduation, Aristotle described common functions, attributes, and assemblages (bone makeup). He stated that some groups, such as humans, were not differentiated into subordinate groups. His ideas on classification lasted until the 19th century, more than 2000 years. He had recognized that nature could be ordered into a ladder of living things from simple to complex. He was a teleologist, believing that all in nature had a purpose and argued against chance growth.

Theophrastos, like his teacher Aristotle, was a thinker. He wrote botanical works that were considered the most important until modern times. He classified trees, shrubs, herbs, the wild and the cultivated, plants of the water and of the land, those with flowers, those with fruit and those without, the evergreen and the deciduous. These distinctions are used even though they may be modified in different areas under different environmental conditions. Theophrastus developed standard botanical language, some still used today. He wrote that plants have the power of generation in all of their parts, because they have life throughout their structure. He knew that plants that appeared to grow spontaneously had grown, because a seed had been carried to that location by wind or some other means. After his death, his works were lost until the 15th century.

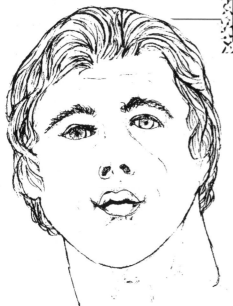

Alexander the Great defeated Persia in 323 B.C. and then sent his general to the throne of Alexandria, Egypt. The city became the intellectual capital of the Mediterranean world. Many races and nationalities visited the great museum at Alexandria. It is believed that Aristotle's Lyceum lead to the organization of the museum. A generation after Alexander conquered the Near East, this center for scientific research was probably begun (290 B.C.) under the first king of Macedonia, Ptolemy Soter (r. 305-284 B.C.). During the reign of the second Ptolemy, Philadelphos (r. 285-246 B.C.), the museum had 100 members. The museum collected species of plants and animals, but it is doubtful that there were any public exhibits.

The library at Alexandria was associated with the museum. It was the largest and is still considered the most famous library the world has ever seen. Library scholars of today still speak of its collections containing almost 500,000 books.

H erophilus made an advance in anatomical description by describing the nervous, vascular, digestive and osseous (skeletal) systems. He distinguished the cerebrum from the cerebellum in the brain, distinguished arteries from veins, and discovered and named the first part of the small intestines-- the duodenum. He showed for the first time that the arteries contained blood. But he also said they carried pneuma or spirit. His student, Erasistratus, later continued his teacher's work in anatomy and physiology.

Erasistratus showed that the epiglottis prevented food and drink from entering the wind pipe during swallowing and noted that food swallowed was ground in the belly. He was the first before Galen to explain the blood systems with pneumas or spirit and air.

Later, at Alexandria, Plato's followers—the dogmatists— claimed there were hidden causes that influenced disease. They had a medical theory that was based on reason and experimental testing. Herophilus and Erasistratus had performed autopsies on men while they were alive. These men were probably criminals given to them by the king. Most of their tests were to prove their theories correct.

Empiricists, on the other hand, followed the Hippocratic beliefs and accepted evident causes. They said that nature was not to be understood except by observation, experience and prognoses. Simple universal causes and theoretical medicine drove these men.

Various medical sects developed and Alexandria's rival center, Pergamon after 250 B.C., developed pergamon (parchment), a material derived from animal skin. As Alexandria's greatest rival they were forced to do so because the Ptolemaic Pharaoh of Alexandria would not allow papyrus or its products to be exported.

Ptolemy had made a regulation stating that all books brought into Alexandria were to be copied and deposited in the library. This library had hundreds of thousands of papyrus rolls, but it was partially burned in A.D. 269. Later, in 414 A.D., it was attacked by an angry mob who wanted the pagan learning to stop. The bishop of Alexandria was the instigator who also assisted in the murder of the woman, mathematician and philosopher who was then head of the museum. The library was burned that day and again during the Islamic invasions. By that time no one was left to copy the writings, and the scholars had taken the valuable manuscripts that were saved. Later, Islam would recover and revive Greek learning in Medieval Europe.

ROME

Greek medicine reached its peak in Alexandria and began to filter into Rome. The Greco-Roman period lasted from 100B.C. to A.D. 600. Rome began its rule over the Greek world sometime after 146 B.C. The Roman version of Greek science was more religiously influenced in medicine. Much of their scientific practices had come from the Etruscans. Copies of Greek sculpture and writing show us that as Rome began to dominate the Greek world, Rome also consumed Greek culture. This Roman version of Greek science was the new energy source of the Latin west in the early Middle Ages.

Asclepiades of Bithynia (120-70 B.C.) was influenced by the teachings of Erasistratus, and he in turn, influenced the acceptance of the Greek practitioners. It is also reported that he restored a dead man to life. He taught physicians that nature cures disease and, for a time, he eliminated the doctrine of the four humors. He improved the standing of the medical doctor in Rome through his practices but was forgotten until the Renaissance.

$\mathcal{M}$edieval scientists were greatly influenced by a poem written by Lucretius (c.95 - 55 B.C.) that presented a complicated psychological theory of sense perception. In his writings on the origin of everything Lucretius said that the world began with the chance assembly of atoms and that life on earth was still progressing. He also presented an evolutionary theory using the idea of natural selection. He believed that selection was made of chance atomic combinations that are better suited to survive while those that are not so well suited die out. His writings did not include Darwin's idea of variation and the descent of species.

A man named Varro wrote an encyclopedia that included medicine. It disappeared early but influenced many during the Middle Ages.

$\mathcal{H}$owever, an understanding of Greek medicine after the first century A.D. can be found in a treatise written by Celsus (fl.c. 25 B.C.). The work was either a paraphrasing of Greek medicine or a direct translation. It shows the understanding of a number of Roman manuscripts (dated to the 9th and 10th centuries) on natural history that were used for agricultural purposes. There is also evidence from that time of an understanding of veterinary science that was found in an excellent treatise written by Ranatus (A.D. 383-450).

Seneca (c.4 B.C. - 65 A.D.) was born in Cordova, Spain. He understood the essentially progressive nature of the development of science and believed that the day would come when the progress of research over the centuries would reveal the hidden mysteries of nature.

PLINY THE ELDER

Caius Pliny (c. 23 A.D. - A.D. 79) worked in public service and wrote many volumes on natural history, 37 of these books were considered the most important collection of books written in Latin. He wrote on 2000 topics that he took from 200 volumes by 100 select authors, all of whom were cited in his texts. He wrote on zoology and botany and described many medical compounds. In the Middle Ages, his writings became quite valuable as a source of information on Greek antiquity. During his lifetime, he gathered as much information he could and died observing the eruption of the volcano, Mt. Vesuvius, that buried Pompeii and Herculaneum.

This man had even stated that
light
travels faster than
sound.

Galen (born in Pergamon, Asia Minor, c. A.D. 129: died A.D. 200) was the greatest medical writer and experimentalist of his time and for 15 centuries thereafter. Medical science and the scientific views on the structure and function of the body would not change until the Renaissance.

He trained in philosophy, mathematics, and natural history and was taught the importance of anatomy as well as the teachings of Hippocrates. His father had pressed him to study medicine because of advice given to him in a dream by Asclepios (the god of healing, c.4th century B.C., in Mycenae when religious and secular practice were one). Galen traveled and studied anatomy in Alexandria. He learned a great deal about surgery during his work as chief physician to the gladiators in his home town. Later he traveled, for the first time, to Rome, where he was received as a great physician. When he tried to leave Rome he was called back by the Emperor Marcus Aurelius.

A prolific writer, Galen wrote in his native Greek. His humoral theory was taken from the earliest of Greek writers. The four humors; phlegm, blood, yellow bile, and black bile, were thought to be responsible for health and disease. Galen classified all personalities on these four humors and described human behavior as being phlegmatic (slow, impassive), sanguine (sturdy, cheerful), choleric (hot tempered), or melancholic (depressed).

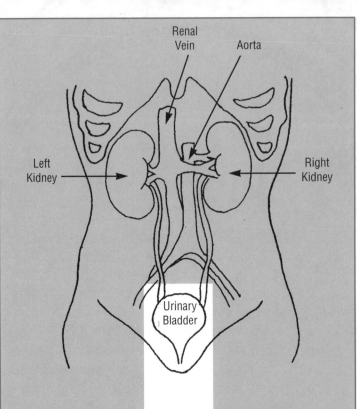

Renal
Vein   Aorta

Left   Right
Kidney →   ← Kidney

Urinary
Bladder

Galen was not a member of any particular medical sect but lectured that Asclepiades and Erasistratus should not have spoken against the assistance of nature, the use of anatomy and the writings of Hippocrates. As medical scientist in the history of early medicine, Galen demonstrated, in dead and in living animals, that urine could never flow backwards. He made many dissections of animals as well as human cadavers, described the great blood and pneuma systems, and classified blood and pneuma as the body's most essential ingredients.

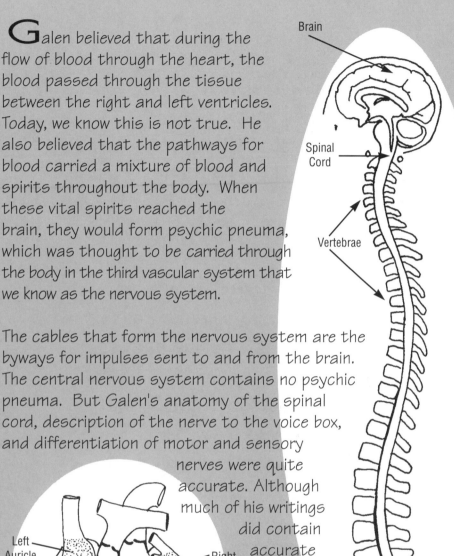

Galen believed that during the flow of blood through the heart, the blood passed through the tissue between the right and left ventricles. Today, we know this is not true. He also believed that the pathways for blood carried a mixture of blood and spirits throughout the body. When these vital spirits reached the brain, they would form psychic pneuma, which was thought to be carried through the body in the third vascular system that we know as the nervous system.

The cables that form the nervous system are the byways for impulses sent to and from the brain. The central nervous system contains no psychic pneuma. But Galen's anatomy of the spinal cord, description of the nerve to the voice box, and differentiation of motor and sensory nerves were quite accurate. Although much of his writings did contain accurate details, from time to time he created data to fit his theories.

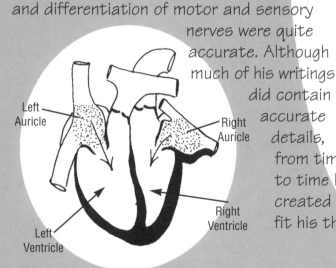

Brain

Spinal Cord

Vertebrae

Left Auricle

Right Auricle

Right Ventricle

Left Ventricle

His representation of Greek medicine (found in the Hippocratic Corpus) was copied and translated during the Middle Ages in Islam (into Arabic) and in the West (from Arabic into Latin). Galen left no question unanswered and his philosophical reasoning and teleological thinking made his writings easy for the christian church to accept. Only 80 out of his 500 texts have survived. For centuries they were commented on and translated. Galen's medical authority remained unchallenged until the anatomist Vesalius began teaching during the Renaissance.

The intermixture of the Greek, Macedonian, Roman, and Near Eastern cultures offered great opportunity for the spread of spiritual movements. The Christian church grew and soon overpowered all other sects that had emerged. As Christianity was accepted as equal to other religions, political persecution of this group slowly ceased. Christianity was seen as a minor Judaic sect after Christ's death. And, not until the Emperor Constantine became a Christian himself was there freedom from persecution. It soon became the official religion of Rome. In A.D. 369, Constantine moved the capital of the Roman Empire to ancient Byzantium and renamed it Constantinople.

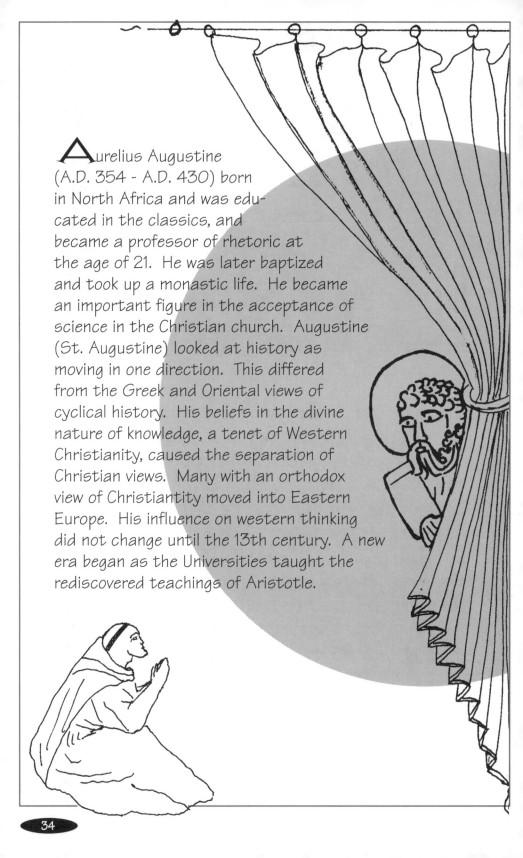

Aurelius Augustine (A.D. 354 - A.D. 430) born in North Africa and was educated in the classics, and became a professor of rhetoric at the age of 21. He was later baptized and took up a monastic life. He became an important figure in the acceptance of science in the Christian church. Augustine (St. Augustine) looked at history as moving in one direction. This differed from the Greek and Oriental views of cyclical history. His beliefs in the divine nature of knowledge, a tenet of Western Christianity, caused the separation of Christian views. Many with an orthodox view of Christiantity moved into Eastern Europe. His influence on western thinking did not change until the 13th century. A new era began as the Universities taught the rediscovered teachings of Aristotle.

Classic scientific authors were re-edited and commented upon until the 5th and 6th centuries. Greek science was converted into Arabic in the 9th and 10th centuries and then from Arabic and Greek into Latin in the 12th and 13th centuries. Arabic and Latin writers and translators made many additions and modifications to the early manuscripts, but Greek learning made its way to the Latin West. By the 13th and 14th centuries, Greek science was again the basis for scientific study. This continued from the Middle Ages into the Renaissance and modern times. It stimulated much of the activity of the scientific revolution in the 16th and 17th centuries. As Seneca had insisted during the Roman era, science would continue to build on its foundations of the past.

Today, science comes from the knowledge received from the Greeks. The foundation of Greek antiquity provided a strong support for the building of all scientific knowledge. So you see, the thinkers of the Renaissance, the scientists and experimentalists, all started where the great scientists of antiquity had left off. This explains the naming of the Dark Ages. It was a time when the human desire for the understanding of the natural world stood still, except in the Near East.

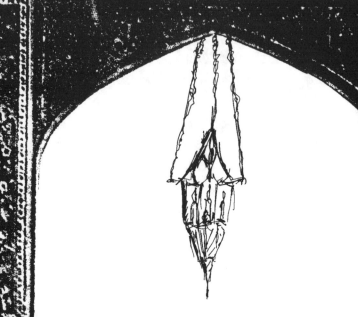

The
Near East

any translations were made by the Arabians. Their knowledge of biology and medicine came from many different sources:- the Greeks, the Persians and the Indians (in India). They studied plants for agricultural and medical purposes.

Abu al-Dinawari (from Iran) in the 9th century and Ikhwan as Safa in the 10th century changed the study of botany, providing details of growth and morphology (structure) that contributed substantially to that science.

Arabian encyclopedias revealed the peoples' interests in animals. The people were familiar and described the habitats and lives of all domestic animals, their reproduction and senses. Their interest in zoology continued into the 14th century.

Around the 10th century many anatomical descriptions were done on the internal organs of the body. But these descriptions were done for religious purposes as in Mesopotamia, where much of their medical knowledge and their practical knowledge of anatomy was inherited from Galen. His works were some of the first Greek texts to be translated and criticized in Arabia. Medicine in the 9th century was basically a revival of Galen's teachings.

Abu Bakr al-Razi (Rhazes, 850-932) was born in Rayy, Iran around A.D. 854 and died around A.D. 930. He attacked religion at a time and in a place where Islam was very strong. He believed that men were equal and did not need religious discipline imposed on them. Religion, to him, was harmful and the cause of war. He felt that science was ever progressive and that it would reach its peak during or just after his time. He spoke highly of men like Hippocrates. But, needless to say, his outspokenness made him very unpopular.

Ibn Sina, the greatest medical practitioner of Islam wrote a great encyclopedic collection on medicine. He described the capillaries and the function of the cardiac valves in the 13th century. His observations of the capillaries were done with the unaided eye, the microscope had not yet been invented. He even wrote on the psychology of animals.

Al-Nafis (c.1210-80) wrote a comprehensive book on the art of medicine while in his thirties and discovered the lesser circulation of blood (the circulation between the heart and lungs). He found the mistake in Galen's observation, that blood flows from the right ventricle of the heart to the left ventricle, three centuries before circulation would be described in Europe. Thirty years after Al-Nafis's ideas reached the West, a European described blood circulation correctly.

THE 11TH CENTURY GREEK REVIVAL

L ike Al-Nafis, many Arabians wrote and commented on Hippocrates and other great thinkers of Greek antiquity. The Greek writings first appeared as Latin translations of Islamic texts. Original Arabian works based on Arabic translations of Aristotle's teachings were circulated.

Commentaries on Aristotle's writings were added to many of these translations. As the translations began to turn up in the West many scholars began to write more translations.

The new universities of Paris (1170) and Oxford were models for our universities of today, the Western European university. Paris became an important city in Western Christian theology and teaching, and Dominicans and Franciscans were teaching at the school in the early 1200s.

At Oxford, Robert Grosseteste made a suggestion for scientific inquiry which contained the essential basis of all experimental science. He believed that man's experience with phenomena was where science began. Science aims to find the causes or reasons for the experience. The causes should be broken down into parts and these principles should then be reconstructed on the basis of a hypothesis (an educated guess). The hypothesis must also be tested, verified or disproved by observation.

Grosseteste's beliefs influenced the thinking of the young Roger Bacon. Bacon had lectured at Paris, returned to Oxford (where he met his teacher), became a Franciscan and was imprisoned because of his belief that philosophy was superior to faith. During his lifetime he revived the concept of the laws of nature.

Albertus Magnus, born in Bavaria around 1200, introduced Greek and Arabic science to the Western European Universities. The church had condemned rediscovered work on natural history written by Aristotle. The ban was lifted in 1234. And Magnus taught Aristotle's works but was outspoken when he did not agree with the Greek philosopher. Magnus believed in observational evidence, the inquiry into nature, and the rejection of myths.

Magnus' treatise describing animals and their development includes insects. He is said to have speculated on the existence of animals in the polar region that would have thick skin and white fur. His study also includes plants. Greek science had begun to stimulate a world of great minds that had been smothered in a culture without a systematic study of nature.

Christian belief and the pagan writings of the philosopher Aristotle were fused by Thomas Aquinas (born c. 1225). He was born to a wealthy Italian family and by the age of 15 went to the new University of Naples. In the 13th century, the scholars of Europe were Dominican Friars. At the age of 20, Thomas Aquinas had joined this religious order and taught that reason alone could not describe the creation of the world. He said that the creation of the world might be a continuous process and was condemned for teaching that the Greek scholars had revealed God's world.

Fifty years after his condemnation and death the Western Catholic church declared him its greatest scholastic model. He was later canonized a saint. Because of Thomas Aquinas and others like him, the physical sciences became the center of focus in the rise of modern science during the Middle Ages.

THE RENAISSANCE

A complete change in the approach to the subjects of nature emerged in the 15th century. During the period between 1500 and 1600, known as the scientific revolution, every science and every method of scientific investigation that had ever been considered was effected. This brought new subdisciplines out of the basic sciences. The studies within natural history and medicine began to intensify.

The change in science came from a general change in European culture, society and the way in which humans (mostly men at this time) viewed themselves and the world in which they lived. This change was known as the Renaissance. It began in Italy in the 14th century with Petrarch (1304-74) and Boccaccio (1313-75). They believed that their culture was heir of classical antiquity. This outlook alone, expressed in poetry, laid the foundation for a new view of life. This new view of life was influenced by the emergence of classical literature. The medieval world began to crumble.

A

new age
of geographical exploration
had begun.
New sea routes and a new
world had been discovered

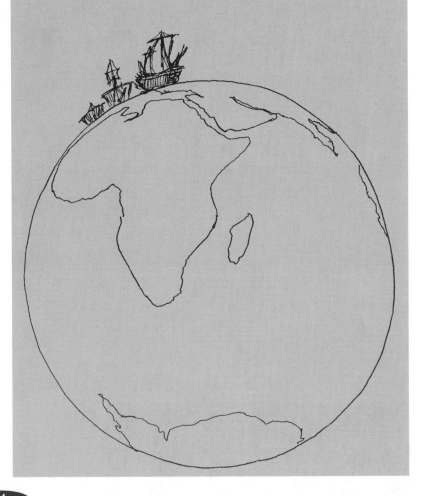

The invention of the printing press (in the mid 1400s in Germany) by Johannes Gutenberg (German, c. 1394-1468) allowed the new information to be printed in books. The books were set up from a stock of separate alphabetical letters. The invention and its process spread quickly throughout Europe. A technique of using engraved metal plates to print illustrations was invented by other goldsmiths and a new group of scholars that printed their works and illustrations began to grow. Many artists began to use the engraving methods as well.

43

lbrect Durer (1471-1528) from Nuremberg, Germany, made engravings of his art and began to print his own books. Durer is famous for his biological drawings of observed specimens. Scholars began to use discovery to create an ordered picture of the universe. They wanted to reveal and understand more of God's creation.

Because the church had difficulties in keeping with the new ideas of the Renaissance, it adopted and incorporated many pagan customs and ideas in order to survive and grow. The early church had begun with good intentions, and over the centuries had become corrupt. Christendom was soon shaken by the Reformation. Protestantism was born and spread through Northern Europe. Protestantism in Durer's part of the world adopted a different attitude towards work that stimulated scientific research and the growth of capitalism.

L eonardo da Vinci (1452-1519)

was born in Tuscany, Italy, and typifies in his broad interests and humanitarian outlook , the pursuits of a Renaissance man.

Leonardo's father recognized his talents when he was young and arranged an apprenticeship with Andrea de Verrochio in his studio. Verrochio was one of Florence's leading artists and received a wide variety of work. At the studio, Leonardo was informally educated and learned perspective. Verrochio was also a goldsmith and sculptor, so it is thought that at his studio Leonardo's mechanical genius was stimulated. Because Leonardo had no university education and none of the experiences of the literary he never published any of his work.

Leonardo kept his ideas until he was commissioned to do something on which he could use them. It is thought that his helicopter design was extracted from his study of bird flight. He never exploited his ideas and was someone who simply wanted to observe the world. He studied the reflection and refraction of light through the eye.

As an artist he needed to become familiar with living forms, which led him to study animal anatomy and to dissect over thirty human cadavers. Despite the unpleasantness of the work, he drew all he saw in great detail.

H is view and drawing of the fetus in utero was one of the first times, if not the very first time, that anyone would see how a human develops. He made no attempt to tell anyone what he learned, and most of his knowledge remained a secret until the recent publication of his drawings.

He became well known in his time for his mechanical abilities and, in 1482, was appointed inspector of fortifications to the Duke of Milan. Leonardo's importance in the history of biology does not lie within the realm of those whose published works have helped the growth of biological knowledge. He represents the changing world during the Renaissance, the scientific revolution, and the beginning of modern scientific thought.

The exact recording of phenomena observed had become the most important procedure in science.

CLASSIFICATION

During the Renaissance, Europeans began to discover the rest of the world. Among these discoveries were strange animals, plants, and different peoples never before seen by Europeans. The work begun by Aristotle was revived and used in a more academic way. Scholars collected plant and animal specimens and wrote short but thorough papers on these new species.

In order to classify things we group together those that are alike and put those that are different into other groups. Different degrees of similarity are taken into account which makes this task somewhat difficult.

An example of classifying or establishing groups might look something like this:

Organisms (everything that is alive): flowers, worms, fish, frogs, cats, humans.

Animals: no flowers, but includes worms, fish, frogs, cats, humans.

Tetrapods (four legged vertebrates): no fish, but includes frogs, cats, humans.

Mammals: no frogs but includes cats and humans.

Primates: no cats but includes humans.

This is a hierarchical arrangement. There are many possible hierarchies producible.

The classification of organisms, still in use today, dates from the middle of the 18th century. Ray (1627-1705) was the English botanist who greatly influenced the making of a scientific system of classification for plants and animals.

CAROLUS LINNAEUS (1707-78)

The father of the modern scientific system of classification.

Born in Sweden, the son of a pastor, Carolus Linnaeus became a botanist at the University of Uppsala. He formalized the use of the hierarchical system of naming plants and animals, and this is the modern system of classification we use today. His book of classification, *Systema Naturae*, appeared in 1735 and went through many editions.

It was once thought that species were fixed unchanging entities created by God. Linnaeus had classified plants and animals for the glory of God, because few biologists at that time believed that the evolution of a new species was possible. Systematists described the species they found as single units created by God. Linnaeus, however, listed 4,235 species of animals. Each was categorized because of characters created by God. And at that time differences or variations within species were explained as being accidental.

Linnaeus's *Systema Naturae* stressed the use of morphological characters (structural characters) to arrange specimens in collections; bones, skin, feathers, leaf shape. But the arrangement was arbitrary and left out the one principle that linked all organismic units, evolution. Linnaeus divided the animal kingdom down to the species that was given a distinct name. But his knowledge of animals was very limited. Today much of his hierarchical system has been expanded and much of his classification altered.

Organisms are arranged in a series of groups. The groups are arranged from less complex to more complex forms of life. Each group is in an ascending order of ever-increasing inclusiveness. This is the hierarchical system of classification. Its major categories, taxa or names, that we use today are: Kingdom, Phylum, Class, Order, Family, Genus, Species. Some of these groups can be further divided into smaller refined groups. More than 30 taxa are recognized.

he binomial system of naming uses two Latinized names and gives each organism a generic name (whose first letter is capitalized) and a specific name (in small letters). Each of these names are italicized (underlined if typed or written). Species that are divided into a subspecies category have a third name. This is called trinomial nomenclature (naming). Only the species level uses the binomial nomenclature. All ranks above the species are uninomial nouns written with capital letters. For example, this is the classification for modern humans: Kingdom, Animal; Phylum, Chordata; Class, Mammal; Order, Primate; Family, Hominidae; Genus, Homo; Species, sapiens.

Before Darwin, each species was believed to be a fixed entity. Each was represented with a type or original specimen in museum collections. Later in 1969, Ernst Mayr said that species should be thought of as groups of interbreeding natural populations that are reproductively isolated from other groups. This means that they can only breed within their group. This concept, the biological species concept, reveals that species are populations in which every individual is unique. Individuals that have a common descent, share integrating characteristics, and have an interbreeding population are considered a species. The type specimen placed in museums is considered an average of the specific group. The biological species concept does not apply to asexual organisms.

No.24. Museum
Nautilus
McCord

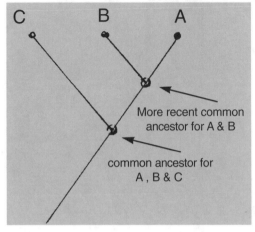

More recent common ancestor for A & B

common ancestor for A, B & C

Cladistics uses a cladogram or phylogenetic tree revealing an evolutionary history. It represents the branching off of related groups of animals from their common ancestors. Branches closest together at the top of the cladogram represent closer evolutionary (phylogenetic) relationships. In order to study the evolutionary relationships of certain groups of animals like a genus (e.g., Homo, humans) within a family (Hominidae) we select another species that we think is related (e.g., Gorilla). The group Gorilla that we think is related is now called the outgroup. Characteristics are chosen that are used to compare a closer related group (Pan, chimpanzees are more closely related to humans than to gorillas) with our chosen species (Homo) and the outgroup (Gorilla). The characteristics may be morphological, physiological, biochemical, behavioral or other characteristics that would change during the course of evolution.

Any of the outgroup (Gorilla) characteristics shared with the related group (Pan) characteristics are considered primitive. These are characteristics possessed by the common ancestor of both Gorilla and Pan (chimpanzee). The characteristics present in Pan but not present in Gorilla are called derived characteristics. Shared derived characters between Pan and Homo would be called synapomorphies. The more derived characters that organisms share, the more closely they are related by a common ancestor.

5 Metatarsal bones   12 ribs   Feathers   Wings   lost orbital bow

Evolutionary systematics bases taxonomy on evolutionary theory. This evolutionary system of classification has been considered the traditional approach to classification since Darwin. It uses many of the same methods of cladistics but uses ancestor-descendant relationships that are not used in phenetics and cladistic classification. Evolutionary systematists use phylogenetic (evolutionary history) trees, with lines that represent the evolution of a species and branches that represent speciation events. These trees show patterns of evolutionary descent.

GORILLA

CHIIMP

MAN

The phylogeny of an animal group represents our concept of the path its evolution has taken. It is the evolutionary history of the group. Evolutionary origin of most animal Phyla is obscure because of Precambrian times (c. 4,600,000,000 b.p.). Fossil records for this time are fragmentary. Since the reconstruction of evolutionary patterns and history are based on comparative morphology and embryology the missing pieces cannot be compared. Research in behavior, comparative biochemistry, serology, cytology, genetic homology, comparative physiology, neuroscience, and other disciplines offer answers to phylogenetic relationships and evolution.

In 1969, R.H. Whittaker proposed a five-kingdom system that uses the prokaryote (single cell) - eukaryote (multicellular) organism distinction.

The Kingdom Monera contains the prokaryotes.

PROKARYOTES

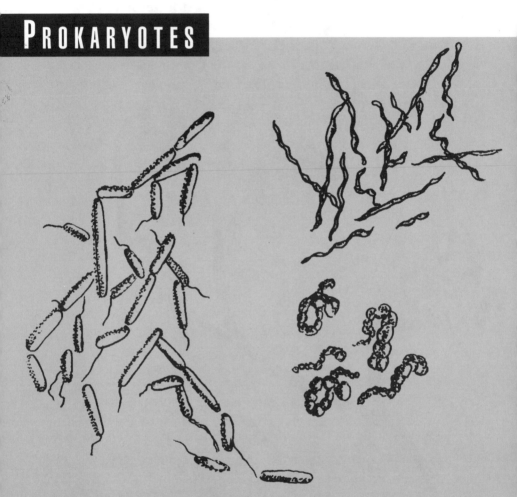

The earliest bacteria-like organisms were very successful, giving rise to a great variety of bacterial forms, some capable of photosynthesis. From these arose the oxygen-evolving blue-green algae (cyanobacteria) some 3 billion years ago. Bacteria are called prokaryotes meaning literally "before the nucleus." They contain a single chromosome comprised of a single large molecule of DNA located in a membrane-bound nucleus but found in a nuclear region or nucleoid.

The Eukaryotes are divided among the remaining four kingdoms.

EUKARYOTES

About 1.5 million years ago, after the accumulation of an oxygen-rich atmosphere, organisms with nuclei appeared. Eukaryotes have cells with membrane-bound nuclei containing chromosomes comprised of chromatin. Chromatin is composed of proteins called histones, RNA and DNA. Eukaryotes are usually larger than prokaryote cells and contain much more DNA and divide by some form of mitosis. Within the eukaryote cell are many membrane-bound organelles, including mitochondria in which the enzymes for oxidative metabolism are packaged. Protozoa, fungi, green and other algae, higher plants, and multicellular animals are composed of eukaryotic cells.

The Kingdom Protista contains unicellular eukaryotic organisms (protozoa and unicellular eukaryotic algae).

The hierarchical system of classification is often changing. Scientists of different disciplines tend to group the organisms in ways in which their discipline orders them. The protozoa were traditionally included in the Kingdom Animalia and within the five kingdom classification divided into seven groups. Because they share many animal characteristics, they are often treated as animals in many text books.

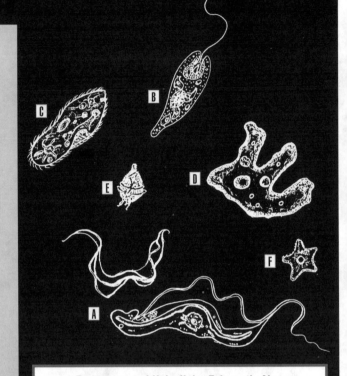

Protozoans and Unicellular Eukaryotic Algae
Classification according to locomotion (movement)

A Trypanosome Invection (insects may act as vectors)
Flagellated Protozoans that move by means of flagella.
Zooflagellates

B Unicillular Algae
Euglena - animal and plantlike characteristic
Phylum Euglenophyta: Euglenoids all have chloroplast and carry out photosynthesis.

C Paramecium
Single cell, most complex of protozoans.

D Amoeba

E Dinoflagellates
Dinophyta: phylum pyrrophyta (seen with scanning electron microscope).

F Golden-Brown Algae - Phylum Chysophyta
Humans can get paralyzed once they have eaten shellfish that have fed upon a group called Marine planktonic algae.

KINGDOM FUNGI

- molds, yeasts, fungi - obtain food by absorption.

Cap

The multicellular organisms are split into three remaining kingdoms:

- multicellular photosynthesizing organisms, higher plants, and multi-cellular algae.

KINGDOM ANIMALIA

- the invertebrates (except protozoa) and vertebrates. Most ingest their food and digest it internally, although some parasitic forms are absorptive.

EVOLUTIONARY BIOLOGY

All biological entities are products of organic evolution. Before the 19th century it was not believed that a new species of plant or animal could develop from an already existing species. Though the idea of evolution was not new to many of Darwin's contemporaries, the cause for evolutionary change was not as yet evident. Charles Robert Darwin and Alfred Russell Wallace explained evolutionary change with the principle of natural selection. The discovery would change the course of modern biology forever.

During the Middle Ages in Europe, Christians believed the earth was fixed at the center of the universe. This universe was made up of spheres that revolved around the earth with perfect regularity. The inorganic and organic matter of our earth was also thought to be fixed and unchanging. There was no concept of evolution, and so the thought of one life form or species evolving was inconceivable. The Old Testament said that living things had been created in order from simple to complex with humans, the most complex, being created last. This ordered picture of life was easily understandable. All organisms were considered linked to one another in a Great Chain of Being. People believed that everything had been created at one time not so long before them. Therefore, no new creatures had come from those before them. They had no idea that many creatures before them had become extinct.

The scholar Archbishop James Usher (1581-1656) worked out what he thought was the exact date of the creation by analyzing the chapter in the Bible that describes the creation. He then dated the creation at 4004 B.C. Quite a short time for the task that others would soon discover took hundreds of millions of years.

New ways of thinking came with the new views of the universe during the Renaissance. The universe was now described as moving changing and heliocentric by Nicholas Copernicus (Polish, 1473-1543), Tycho Brah (Danish, 1546-1601), Galileo Galilei (Italian, 1564-1642) and Isaac Newton (English, 1642-1727). Some of the most famous men during the renaissance and well into the 19th century were driven by the human duty to study the natural relationships of things in order to demonstrate the glory of god. This same conviction lead Darwin to his discovery of the principle of natural selection. All of Darwin's professors at the University of Cambridge were churchmen. "...The radical change of the intellectual climate of the seventeenth century enabled scientists of the following century to think about the universe..." in a way "...that made the idea of evolution possible.... (Nelson and Jurmain, p. 30, 1991)."

his Systema Naturae. And though this went against the belief that God made man (almost all scholars at this time were men, hence the misogynist views) in his own image, and that he should be considered separate from animals, the system was received well throughout Europe.

Linnaeus thought and wrote that every living thing is created perfectly, adapted to the environment in which it lives. This, of course, would make change unnecessary. Many scientists had already given up the idea that nature was fixed. But because Linnaeus believed all organisms were divinely created, he studied and diligently assigned every living thing its proper name for the glory of God. To him, species could not mutate, and there could be no new species. Other scholars were in favor of a dynamic or changing universe. These scholars believed that the relationship between similar organisms was based on descent from a common ancestor.

I n the previous section we introduced Carolus Linnaeus and his ideas for the classification of plants and animals in Systema Naturae. Like the other scholars of his time (18th century) Linnaeus believed that all of the animals created at the beginning were still living. He did not believe in new species and agreed with the idea of a Great Chain of Being. He included humans in

During the same time, a man known simply as, Buffon (1707-88) (once Georges Louis Leclerc from Burgundy, France, later, made a count by the king under the name of Buffon) whose volumes entitled Natural History were best sellers, disagreed with Linnaeus on most points. The volumes influenced many of the leading scholars of Europe and made natural science a popular conversation topic at parties and in the drawing-rooms of Europe. Buffon was a scientist whose revolutionary ideas would not be accepted for another generation. He recognized the dynamic relationship between environment and nature. Buffon believed that if a group or species was to be separated, by migrating, into different areas of the world a new form or species could develop. The new species would develop, because each separated group would be influenced by their local climate conditions. He rejected the idea that one species could turn into another species. This was known as transformation. But, he did believe in the mutability of nature and worked out a scheme for the evolution of the earth. This was similar to Usher's work but he wrote that the earth was 76,000 years old (it was quoted that in private he said millions of years).

Nature, he felt, could be described by a system of laws, elements, and forces. The dynamic system of life had processes that should be discovered and understood separately from theology and the belief of a divine creator. There were various and minute gradations of nature. Buffon actually managed to mention every important idea that would be incorporated in Darwin's *The Origin of the Species*, but he did not piece the information together as Darwin would. At the end of the 18th century, though the idea of species being mutable and the possibility of unlimited organic change was not widely accepted, it was being argued and discussed a great deal in intellectual circles. These ideas were well published and discussed throughout Europe, and it is well known that Darwin was familiar with Buffon's voluminous publications.

J ean Baptiste Pierre Antoine de Monet de Lamarck of France (1744-1829) was the first European scientist to create a comprehensive system of the natural world and was a major proponent of the idea of organic evolution. He created a system with three points that explained the transformation of one species into another. In this system, he explained the spontaneous generations of the simplest life forms, the evolution of these simple forms into more complex ones, and mechanisms within the environment that slowed the natural world's tendency toward complexity. And, in this system, new species were explained as evolving because of the use or disuse of organs. He believed that when an animal changed habitat, the disuse of an organ would cause its disappearance. On the other hand, frequent use would cause its development. He felt that if some sort of primate or tree dwelling animal were to leave the trees, and, generation by generation, begin to walk, they would become bipedal, their thumbs on their feet (big toe) would eventually look like ours. He believed that an animal could propel its own change and that its habits would change, the fluid's pathways, eventually changing the body's form.

There was no evidence supporting spontaneous generation, trends toward complexity, and the actions of fluids; nor could Lamark's notion of the inheritance of acquired traits be scientifically supported. But his system was so well accepted in the scientific community that he did not feel the need to prove it. Darwin would use this belief that acquired characteristics could be inherited. Lamark was the first to work out the theory of descent as an independent scientific theory and the philosophical foundation of all of biology (Clodd, 1897, p.115) In fact, Lamark popularized the idea of evolution. Soon Darwin would explain evolution and inheritance much more adequately with the principle of natural selection.

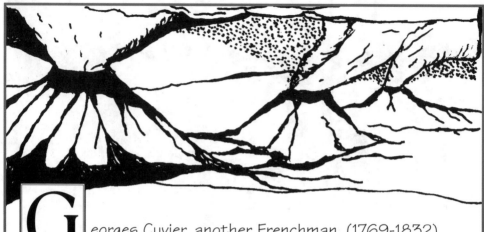

G eorges Cuvier, another Frenchman, (1769-1832) was the father of zoology, paleontology, and comparative anatomy. Cuvier was the most outspoken opponent of evolution during that period. He insisted upon the fixity of species. Nevertheless, he separated his religious beliefs from his scientific work. He proposed his theory of catastrophism to destroy growing interest in the theory of evolution. He taught that sudden catastrophes had destroyed life at different times on earth, and that the organisms from neighboring areas replaced those that had died.

Cuvier advanced the classification of the animal kingdom by separating vertebrates and invertebrates into four divisions using anatomical and physiological characters. The practice of classification would soon show scientists that species were capable of change, and that life was not linear. Science had come closer to correcting the belief of a Great Chain of Being.

Darwin's grandfather, Erasmus Darwin (1731-1802) was a country doctor, poet, and scientist. His contribution to organic, including human, evolution was one of the greatest. It is said that the theory of natural selection was the only principle idea in the evolutionary system that Darwin's grandfather did not pass on to him.

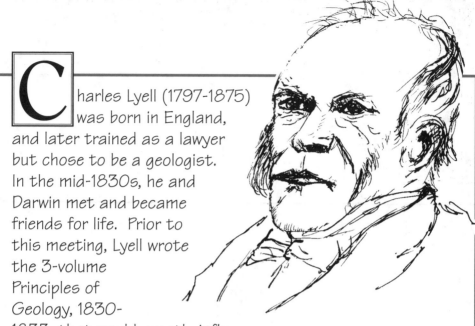

harles Lyell (1797-1875) was born in England, and later trained as a lawyer but chose to be a geologist. In the mid-1830s, he and Darwin met and became friends for life. Prior to this meeting, Lyell wrote the 3-volume Principles of Geology, 1830-1833, that would greatly influence Darwin's thoughts on the mechanisms of evolution.

Lyell supported the principle of uniformitarianism that James Hutton had introduced in 1785. The principle stated that all forces active in the earth's past are working today. It described the earth's crust as being formed through a series of slow gradual changes and maintained that all of nature was the result of natural forces. He wrote that these forces, like erosion, could have caused all the geological events in the past as well as those in the future.

The Beagle

At the beginning of his scientific expedition the branch of natural history in which Charles Darwin (1809-82) was most interested was geology. And it was during this five-year trip that he received the second volume of Lyell's Principles of Geology. Lyell wrote on the vast age of the earth and the development of the earth's many surfaces. He also wrote that the struggle for existence could change living forms. Darwin credits Lyell with his use of the principle of struggle for existence. Darwin's idea of decent with the alteration of an organisms morphology was described as a slow process that was gradual. This scheme fit perfectly within Lyell's belief that the earth was hundreds or thousands of million years old.

The clergyman and botanist John Stevens Henslow, who took Darwin on long plant-collecting expeditions, invited him to take up a post as a naturalist on one of His Majesty's surveying vessels. On September 5, Darwin was interviewed by the Captain FitzRoy of the HMS Beagle. After many delays, the ship sailed on December 27, 1831 and did not return to England for another five years. As a creationist, Captain Fitzroy believed that slavery was the natural order of things. Locked together for five years, Darwin and Fitzroy represented the two confronting but major thoughts on the natural world.

During his trip, Darwin kept a journal of his daily investigations and of the trip itself. On occasion, Darwin left the ship to make long overland excursions in South America. He collected hundreds of specimens, dissected and stuffed them and examined the geologic scenery as well. This map shows the course which Darwin and the Beagle took around the world surveying the coastal waters.

During his life, Darwin collected a great amount of material in support of the theory of evolution by natural selection. Darwin knew that breeders had greatly modified varieties of domestic species during historic times through the artificial selection of plants and animals. This information provided strong support for the process of natural selection. Before his trip on the Beagle, Darwin had collected a great deal of information on the flora and fauna of his native England. With this information and the information collected during his study of the world on the Beagle voyage, Darwin had realized that isolation was an important mechanism within natural selection. Darwin found that in small isolated or confined areas, organic and inorganic conditions of life will generally be uniform. It was clear to Darwin that all of the individual organisms within a variety of species in an area are influenced and modified by the same environmental pressures.

Darwin knew that the evidence of a slow and gradual evolutionary change would come from a record of the fossils (the fossil record) in the earths layers (strata). Though he realized the fossil record could never be complete, it strengthened his argument, provided new directions, and greatly encouraged future research.

The English political economist and clergyman Thomas Robert Malthus (1766-1834) was also devoted to the natural sciences. He is famous for his research on population problems. In an essay on the principle of population, he pointed out that the human population would double every year if it were unrestrained by natural causes. He showed that food production increased in a straight arithmetic progression. In nature, food production is checked by the struggle for existence. Today, our society applies restraints to increase food production. Malthus also explained that humans were infinitely fertile, but that the limited size and number of resources on the earth kept the population from growing out of control.

The circumstances Malthus presented helped Darwin realize there was a selective process taking place, and that selection in nature could be explained as a survival of the fittest. It was the individual (not the species) with favorable variations that would survive. Those with unfavorable variations would not. Individuals interacting with one another in a struggle for survival led Darwin to his concept of natural selection.

Today, Ernst Mayr of Harvard University has said that there is no typical individual. We have come to understand natural selection as a force or mechanism that acts on individuals. These individuals survive or perish and the population evolves in a certain direction. The population that evolves is made up of individuals that have been "selected for" because of environmental conditions in the areas in which they live.

Alfred Russell Wallace

Darwin allowed the idea of selection to develop in his head and wrote an essay on natural selection in 1849. Darwin did not publish the essay. He had planned to detail all of the information in a multi-volume work. In 1856, he began to put his voluminous data into a work on the origin of species. His plans to write four volumes changed abruptly when he received a manuscript from Alfred Russell Wallace in 1858.

Alfred Wallace was an English naturalist living in Malay with whom Darwin had correspondence. Wallace's paper "On the Tendency of Varieties to Depart Indefinitely from the Original Type" (1858) made Darwin quite nervous. Darwin wrote his friend Charles Lyell, who had pressed him for years to publish his idea. Lyell helped decide that joint papers by Wallace and Darwin should be read at a meeting of the Linnaean Society. Darwin's summary of his work in the The Origin of Species, and Wallace's paper were read on the evening of July 1, 1858. Neither Darwin nor Wallace attended the reading. Together these scientists provided the first credible explanation for evolutionary change.

Evolution is the underlying principle in the biological sciences. But the evidence for evolution lies within the boundaries of many disciplines.

In geology, evolution is a dynamic system that is continual and has progressed for millennia. Animals living today can be linked with one another by extinct common ancestors found buried in the earth. The extinct forms are usually different from today's living forms. The discovery may be the fossil ancestor of two living animals that were once thought to have separate evolutionary histories. If this is so, the discovery allows us to join the two living groups into a type of family tree, and the evolutionary history of the two animals becomes a little clearer.

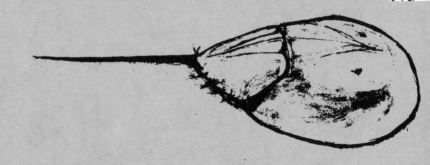

W hen we compare the anatomy of animals in the same group we see likenesses in the general organization of their structures. These likenesses are independent of other animals in their environments or habitats. Some of the structure and organs of completely different species that are grouped together will be homologous or alike. In 1859, Darwin said that morphological study, the study of anatomical structure, shows that the human hand is made for grasping, the whale's fin for paddling, the bird's and the bat's wings for flying. And all are constructed on the same pattern with the same bones in the same relative positions.

We know that all of the animals living on the earth have evolved from animals living in the past. When we find a number of animals that have many morphological or structural characters in common, we know they share a common descent. The more characters the animals have in common, the more closely related they are which also makes them more closely related to their common ancestor. This is how morphology contributes to the phylogenetic scheme that forms the basis for the modern classification of animals.

A few common characters shared by two organismal groups have little significance because of the possibility of convergent evolution. In other words, similar environmental pressures may have selected for organisms (chosen organisms through natural selection) with similar adaptations in unrelated groups during their evolution. But when common characters are homologous (similar in origin) the evidence for an evolutionary relationship is strong.

Evidence of evolution is in the vestigial organs (i.e., our appendix), which apparently have no function. The theory of evolution explains the presence of rudimentary organs as those that are retained during evolution but have become useless. These useless organs left creationism with a disturbing unsolved mystery. Darwin thought that these organs where inherited and could possibly be predicted and explained through the law of inheritance.

Darwin also recognized that the immature stages during an organism's embryological growth give clues to evolutionary relationships. The American naturalist, and Swiss born, Louis Aggassiz (1807-73), said that the mammal, bird, and reptile embryos he collected could not be sorted at a later date if they were left without labels at the time of their collection. The embryos were too similar.

Embryogenesis or the study of embryonic development of animals in an evolutionary context shows that embryos often go through stages resembling their ancestors. In the biogenetic law it was believed that embryonic stages of an animal were similar to the adult stages of its ancestors. This principle of recapitulation or the idea that ontogeny (life history) repeats phylogeny (evolutionary history) has been shown to be incorrect. But Stephen J. Gould (1977) of Harvard University has shown that early developmental patterns tend to become more or less stabilized in the life history of generations that follow their ancestors (paleogenesis). Scientists have restated the biogenetic law saying that mammalian embryos go through stages that resemble the embryos of their ancestors.

Natural Selection: Mechanism of Evolution

The most famous documented case and modern example of natural selection acting in modern populations

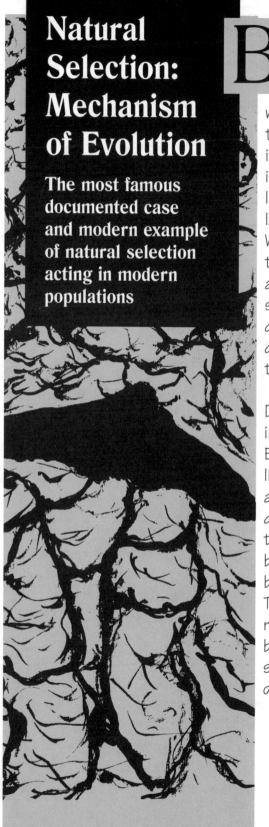

Before the 19th century, the common variety of moth in England was of a peppered or grey color whose surface was much like the lichen covered trees in its habitat. Another moth in the same environment but lower in number in the population were dark moths. When resting on the trees the spotted moth was almost invisible to birds that shared the habitat. The dark moths were seen immediately against the light trees and eaten by the birds.

During the industrialization in the 19th century in England, pollutants killed the lichen on the trees in these areas and turned the bark a dark color. Both moths continued to rest on the trees but the grey moth began to be easily recognized by birds. The dark moth populations rose as the light grey moth became the visible prey, consequently lowering the frequency of their populations.

With the increase in pollution control during the 20th century, some of the forested areas have returned to their preindustrial state, and lichen is again growing on the trees. The dark moth population is now on the decline. The substance that produces pigmentation, melanin, in an organism, and the evolutionary shift in the peppered moth caused by industry is called industrial melanism. This evolutionary change is called adaptation.

All animals evolve by the mechanism of natural selection. A particular characteristic's function helps an organism or a population survive and reproduce. In the sciences, function is a unique entity to biology. Characteristics that help an animal or population to survive and to reproduce are said to have an adaptive value. These characteristics have been selected and become adaptations over the course of evolution.

All organisms are adapted to their habitat. Environmental conditions vary from one habitat to another. The conditions of each habitat contain pressures that favor the survival of organisms with certain characteristics over organisms with characteristics less equipped or adapted to that environment.

In order for a trait to have importance in natural selection, it must be inherited. The grey moths inherited their color (this has been demonstrated as a hereditary trait) from their parents, and this color is an important trait in their survival. As the grey moths breed and pass on the peppered grey color, their population increases. The dark moths still producing dark offspring are recognized against the light colored trees and eaten, which decreases their population.

B ecause there was a variation in the inherited traits of the moths, natural selection could occur. If all the moths were initially grey and the trees became dark, survival and reproduction of all the grey moths may have become low enough to cause extinction. Selection works on variations in populations.

Fitness or reproductive success is a relative measure that changes as the environment changes. The grey moth was most fit during the preindustrial era and the dark moth most fit as England industrialized and the trees became dark. The fitness of an organism depends on the specific environment in which it lives.

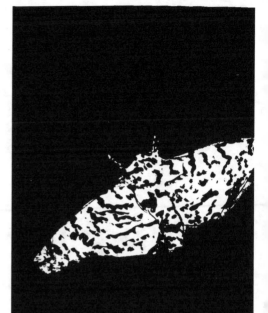

Death rates also influence natural selection. Logically, moths dying early leave fewer offspring. Fertility (or the ability to have offspring) is an important aspect of natural selection. Animals that give birth to more young will pass their genes on at faster rates than those who bear fewer offspring. But the most important element of fertility is the number of young raised successfully to the point where they reproduce themselves.

Fertility or reproductive success can be measured as the number of offspring that live to reproduce. An example of reproductive success for three organisms might be:

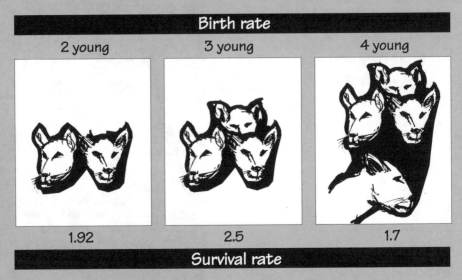

Birth rate		
2 young	3 young	4 young
1.92	2.5	1.7
Survival rate		

Averaged over many generations, the survival rate for those organisms bearing 2, 3, and 4 offspring in a hypothetical habitat are compared. In this hypothetical habitat, we see that the animals having 3 young have a higher reproductive success. In other words, having 3 young is more beneficial than having 2 or even 4.

The number of young born in a litter or the number of eggs laid by birds are products of an evolutionary strategy that may be genetic. The number of offspring allowed to be carried by a mother at one time seems to have been selected for over the course of evolution, so that reproductive success is highest for the individuals within a particular species.

Darwin did not know how to explain variation or the transmission of traits from parents to offspring. In 1859, inheritance was thought of as a blending of traits by nature. But one of Darwin's contemporaries had already systematically worked out the rules of heredity. These were not to be recognized again until the beginning of the 20th century.

T he explanation for the diversity of life on earth is organic evolution. The characteristics and distribution of the organisms present today are explained in their evolutionary histories. Their long but gradual and continuous changes are irreversible (Dollo's law) and have been punctuated by periodic mass extinctions, followed by the rapid spread of survivors into other environments. This theory of punctuated equilibrium, proposed by Gould and Eldridge in 1972 states that new species arise almost instantaneously in geologic time and then remain almost unchanged during the rest of their history on earth.

Fossils found by paleontologists support this theory. The age of the fossils are reliably estimated because the geologic formations where they are found can be dated by measuring the radioactive decay of the naturally occurring elements within them. The fossils found by paleontologists are used by evolutionary morphologists to research homologous morphological characteristics that help reconstruct evolutionary relationships.

Timetable of Evolution

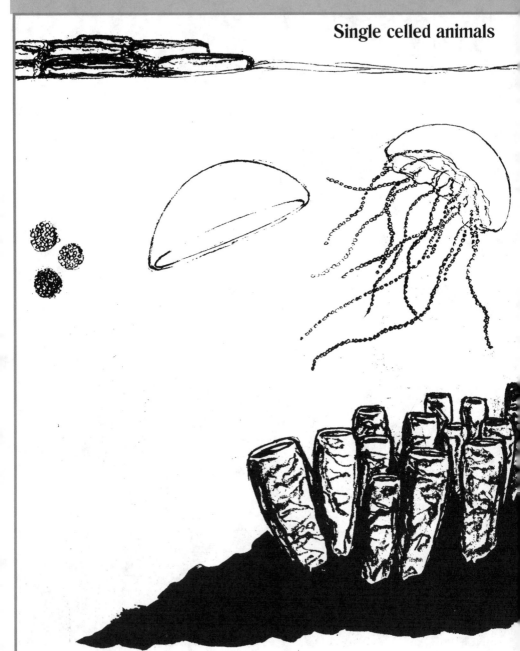

Single celled animals

First jawed fish

Vertebrates move onto the land

Dinosaurs emerge

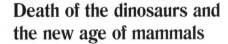

Death of the dinosaurs and the new age of mammals

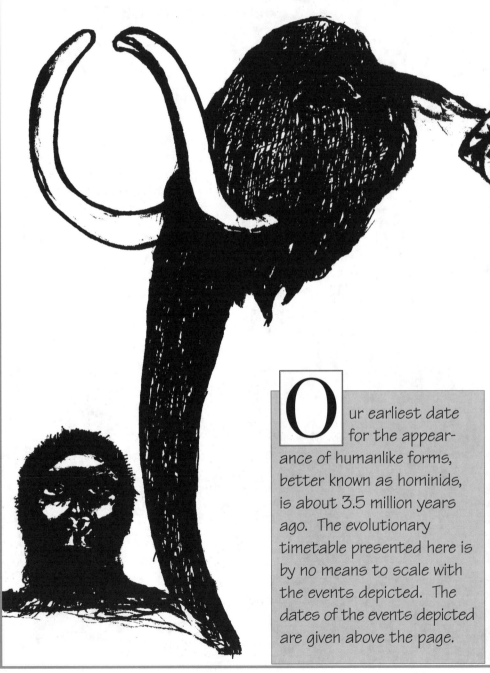

Our earliest date for the appearance of humanlike forms, better known as hominids, is about 3.5 million years ago. The evolutionary timetable presented here is by no means to scale with the events depicted. The dates of the events depicted are given above the page.

Hominids

Basic Human Anatomy and Physiology

ORGANISMAL STRUCTURE AND FUNCTION

The term anatomy, as we know, was first used by Aristotle. We described the rediscovery of Greek and Roman works during the Renaissance and at the beginning of the scientific revolution. Many men like Leonardo da Vinci began to change the way in which we viewed our world. Observing, recording and teaching anatomy was changed forever with the work of one man.

Andrés Vesalius was born in Brussels (1514-64) and was trained in medicine at the Universite de Paris. He completed his studies after the war between France and the Holy Roman Empire at the medical school of Padua. At 28, he taught surgery and anatomy. He revolutionized anatomy and scientific teaching with his *Opus Magnum*. His *De Humani Corporis Fabrica* (1543), one of the world's greatest books, is filled with illustrations that supplemented the *Opus Magnum*. In the Corpis Fabrica, the organ systems were considered separately.

Vesalius recognized that all structures were not identical in all humans. He found many errors in Galen's work but continued to follow Galen's physiology of the heart. The artist of Vesalius' Fabrica is not known. Many scholars think that the style of artistic sophistication and knowledge of the new techniques of the Renaissance are the work of Jan Stephen van Calcar (1499-1546/50).

Clinical anatomy emphasizes the structure and function of the human body as it relates to medicine and the other health sciences. Cadaver anatomy is now the foundation of the knowledge of anatomy. In 1895, the German physicist, Wilhelm Roentgen (1845-1923) discovered x-rays and observations of the living skeleton were made available. Radiopaque and radio translucent substances allow us to study (radiological anatomy) various organs and body cavities. Today's imaging techniques such as ultrasound, computer tomography (CT), and magnetic resonance imaging (MRI) allow anatomical observation in the living.

Learning anatomy is mostly making oneself familiar (memorizing) with terminology and related structures and their function. But this study is a visual one. By using cadavers, diagrams, illustrations and photographs, students can identify every angle of an organism's structure and identify structures of the body in any manner that they are presented.

In anatomy courses the body is studied by region (dissections are done best this way and the cadaver holds up better). When describing function it is best to describe the body in terms of systems. Since we are not using cadavers we will discuss regional structure and its accompanying function together. We must also remember that anatomical terminology introduces us to medical terminology that is mostly derived from Latin and Greek.

The Integumentary System is composed of the skin and its appendages (eg. hair and nails). This is visible for you firsthand.

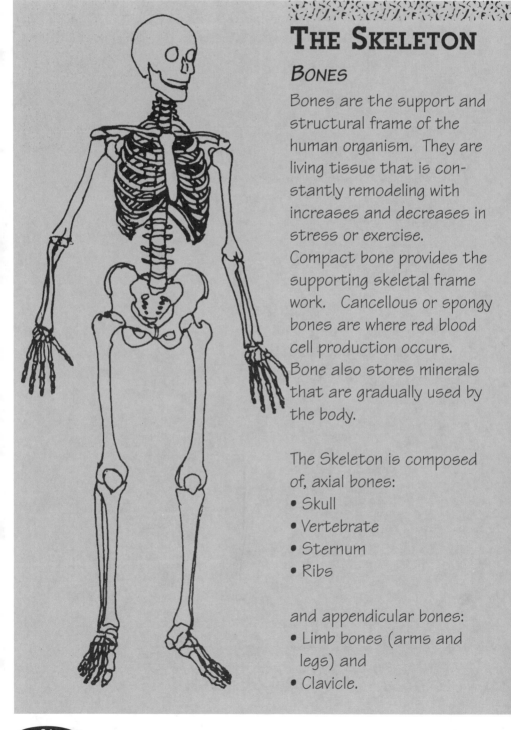

THE SKELETON

BONES

Bones are the support and structural frame of the human organism. They are living tissue that is constantly remodeling with increases and decreases in stress or exercise.

Compact bone provides the supporting skeletal frame work. Cancellous or spongy bones are where red blood cell production occurs. Bone also stores minerals that are gradually used by the body.

The Skeleton is composed of, axial bones:
• Skull
• Vertebrate
• Sternum
• Ribs

and appendicular bones:
• Limb bones (arms and legs) and
• Clavicle.

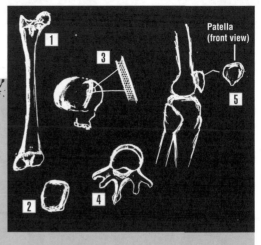

1. Long bones are tubular in shape with two ends that are either concave or convex (leg bones, arm bones, etc.).

2. Short bones are cuboidal in shape and found in the foot or wrist.

3. Flatbones are protective reinforcements (like the flat bones of the skull). They are made of two plates with spongy bone between them.

4. Irregular bones are various shapes and sizes like the facia bones and the vertebrate)

5. Sesamoid bones are round or oval and develop only in certain tendons, like the patella, where tendons cross long bones.
(Accessory bones develop at additional ossification centers and give rise to extra bones. This is common in the foot.)

Living bones are plastic (they have plasticity) tissues containing organic and inorganic compounds. They protect vital structures, support the body and are the mechanical basis for movement. Some of the important functions of longbones are the development of red blood cells and the storage of salts as a mineral reservoir for the body.

JOINTS

Bones meet one another at places we call joints. Some joints are free moving, while others hardly move at all.
A joint's stability when stationary or during movement is aided
by tough fibrous bands or ligaments that surround the joint capsules. There are three types of joint:

- Fibrous joints are united by fibrous tissue.
- Cartilaginous joints are united by cartilage or cartilage and fibrous tissues.
- Synovial joints are united by cartilage with a synovial membrane enclosing a joint cavity.

The knee joint (#5) is highly mobile. The ends of the bones that meet joints such as the knee joint are lined with cartilage that is enclosed with a capsule containing synovial fluid. Synovial fluid is a special lubricant that reduces friction between the two opposing surfaces. Industry has never been able to duplicate a lubricant as effective as the synovial fluid produced by the body.

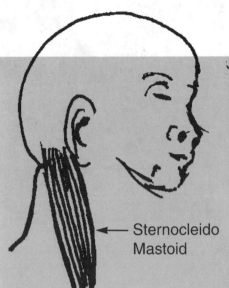

Sternocleido Mastoid

MUSCLE

There are three types of muscle:
1. Skeletal muscles
2. Smooth muscle
3. Cardiac muscle

The sternocleidomastoid helps move the head.

The framework of bones that make up the skeleton and their connecting ligaments are moved at the joint by a network of muscles surrounding the skeletal system. This network of muscles and the skeletal system are sometimes referred to as the musculoskeletal system.

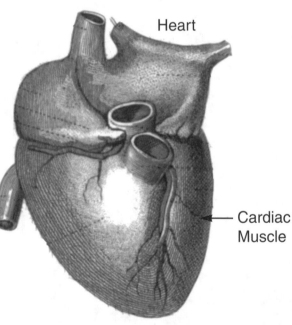

Heart

Cardiac Muscle

Musculoskeletal System

Skeletal Muscle

Joints are moved by skeletal muscle (section of a typical limb muscle) composed of fibers that are compartmentalized by commands from the nerves that ennervate and control them. These bundles of fibers contract and relax so that the muscles shorten and lengthen. With different abilities, they are specifically designed to contract quickly or respond with more precision in a slow but steady movement. Each fiber is proportioned in each muscle to respond accordingly to its function.

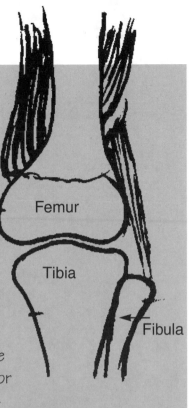

Femur

Tibia

Fibula

The other two types of muscle are cardiac muscle (making up the walls of the heart) and smooth muscle in the walls of most vessels and hollow organs (like the stomach).

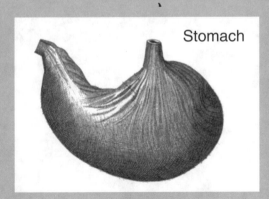

Stomach

THE LYMPHATIC SYSTEM

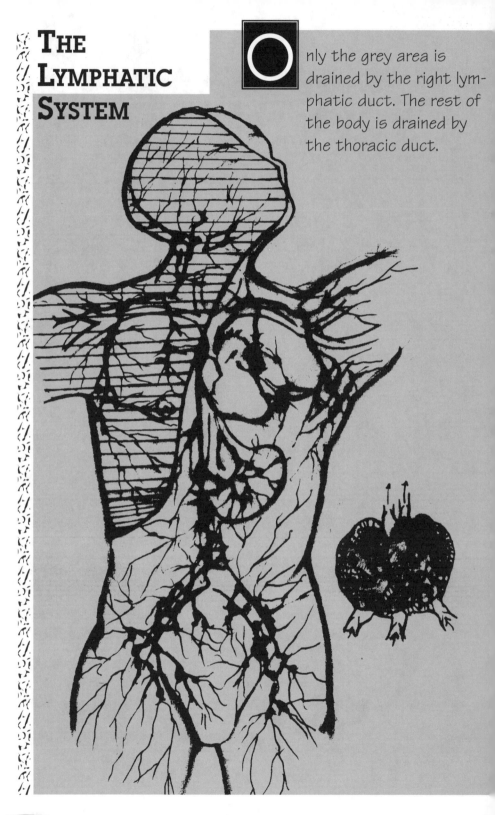

Only the grey area is drained by the right lymphatic duct. The rest of the body is drained by the thoracic duct.

The lymphatic system functions as a collector of unwanted materials from the body's tissues. This unwanted material is filtered through nodes that run into larger ducts and finally drain into the venous system (veins). Because of its major role in fighting foreign bodies, the lymphatic system is often the cause for the spread of cancer and other infectious diseases.

The lymphatic system is divided into two major areas of the body. The lymphatics of the upper right limb enter the axillary lymph nodes in the axilla. The lymphatics of the head and neck drain into the cervical lymph nodes mainly grouped around the vessels of the neck. Most areas are drained by the thoracic duct; the remaining area is drained by the right lymphatic duct. The networks of very small lymph vessels are lymphatic plexuses called lymph capillaries.

Lymph nodes are small masses of lymphatic tissue along the vessels where lymph passes through on its way to the venous system. Aggregations of lymphoid tissue lie in the walls of the alimentary canal (i.e., tonsils), spleen, and thymus. Circulating lymphocytes formed in lymphoid tissue are located throughout the body (i.e., lymph nodes or spleen) and in myeloid tissue located in red bone marrow.

THORAX (CHEST)

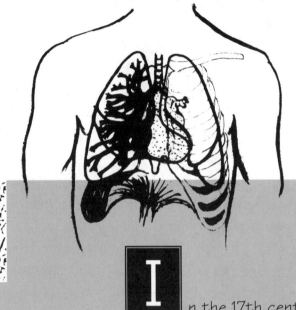

I n the 17th century, Boyle's (1627-91) experiments showed that the combustion of a candle and the life of an animal were sustained by air. Oxygen (O_2) is fundamental in the conversion of food to energy in the body. It is taken from the atmosphere by the respiratory system. Carbon Dioxide (CO_2) is a waste product removed during the conversion from food to energy.

The major muscle of respiration, the diaphragm, lies at the bottom of the rib cage. The diaphragm is innervated and receives messages from the brain through the phrenic nerve. The diaphragm and other muscles contract and enlarge the lung cavity. Atmospheric pressure forces air into the lungs, expanding them. When the diaphragm relaxes, the lungs contract and air is expelled.

THE LUNGS

Air

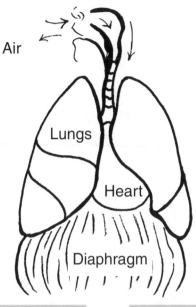

Lungs

Heart

Diaphragm

Air
↑
Relaxing
Diaphragm

Contracting
Diaphragm

Air

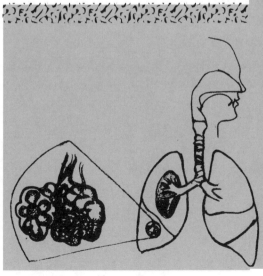

The oxygen taken in by the respiratory system is absorbed into the blood-stream inside the lungs. The lungs are large, spongy air sacs composed of compart-mentalized separate units called *bronchopulmonary segments*. Air travelling through the windpipe is dis-tributed throughout the terminal branches of the bronchial tree and into each segment of the lung. Blue blood (lacking oxygen) travels through the segments, picks up oxygen, and becomes red blood. The carbon dioxide is removed from the blood and replaced by the oxygen needed by the body's tis-sues in tiny thin walled air sacs called *alveoli*.

CIRCULATORY SYSTEM OR VASCULAR SYSTEM

William Harvey

Micheal Servetus (1511-1530) was the first to suggest a separate movement of the blood through the lungs. Gabriello Fallopia (1523-62) was a student of Vesalius. Fallopia (who described the fallopian tubes) taught Fabricius about Aquapendente (1537-1619) who in turn taught William Harvey. Fabricius' description of the valves in the veins was an important observation that Harvey would later use in his own research.

William Harvey (1578-1657), educated at Cambridge proved that the blood continuously circulated through a closed (contained) system. This discovery was one of the most significant achievements in medicine and physiology in the 17th Century. Harvey is responsible for our present understanding of the blood's circulation. He demonstrated that the heart was a pump and that blood moved toward the heart in a great Vena cava and away from the heart through the main artery the aorta. He wrote "On the Movement of the Heart and Blood in Animals."

BLOOD

The bloodstream is the body's transportation system. It transports the oxygen and nutrients the tissues need and disposes of waste products. Hormones travel between glands and tissues and antibodies travel to locations of infection and disease via the bloodstream. The circulatory system regulates internal body temperature as well.

Blood is a complex substance composed of a clear fluid called *plasma* and an assortment of specialized cells. Red cells carry the oxygen; various kinds of white cells fight disease and platelets repair damaged blood vessels. Blood is always opaque and is bright scarlet red when oxygenated and purple when depleted of oxygen.

THE HEART

The heart powers the entire circulatory system, pushing the blood throughout the body. As we mentioned earlier, the heart is composed of smooth muscle. In mammals the heart is divided into four chambers. The thin-walled chambers called *atria* receive incoming blood while the thick-walled ventricles pump it out.

BLOOD FLOW IN THE HEART

Blue or deoxygenated blood flows through the superior and inferior vena cava into the right atrium (on left). The tricuspid valve opens and blood passes to the right ventricle, which contracts, sending the blood through the pulmonary valve and trunk and then into the lungs. Oxygenated blood returns from the lungs via the pulmonary veins and enters the left atrium. The left atrioventricular (bicuspid or mitral) valve then opens, allowing the flow of blood to the left ventricle. The left ventricle contracts, and the fresh blood is finally pumped through the aortic valve and travels through the ascending and descending aortas (the main artery) to the rest of the body's tissues. The system of valves prevents the back flow of blood between the heart's chambers.

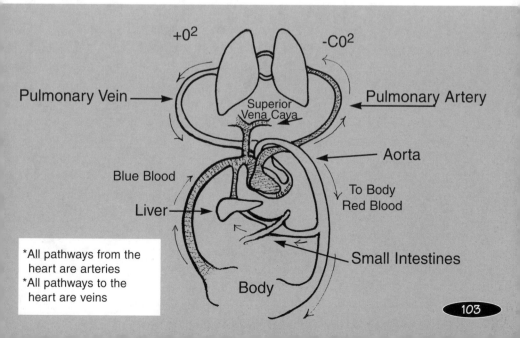

$+O_2$

$-CO_2$

Pulmonary Vein

Pulmonary Artery

Superior Vena Cava

Aorta

Blue Blood

To Body
Red Blood

Liver

Small Intestines

*All pathways from the heart are arteries
*All pathways to the heart are veins

Body

ARTERIAL AND VENOUS SYSTEMS

THE ARTERIES

The blood is forced from the heart and travels through the arteries to the various tissues of the body. As arteries penetrate the tissues, they become smaller and branch out into fine capillary networks that lead into the veins or venous system.

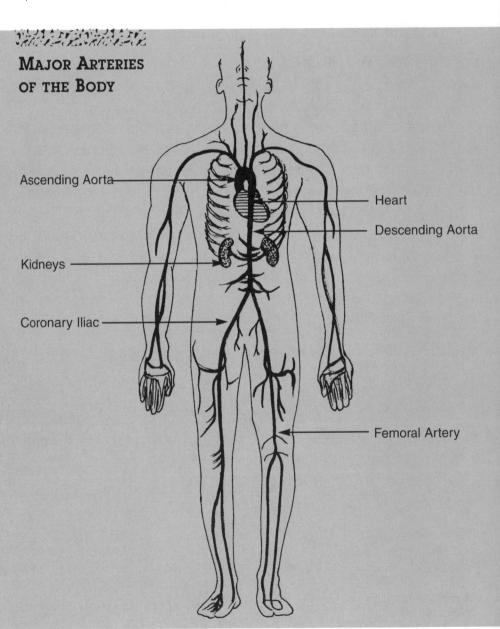

MAJOR ARTERIES OF THE BODY

Ascending Aorta

Heart

Descending Aorta

Kidneys

Coronary Iliac

Femoral Artery

THE VEINS

Veins carry the depleted blood back to the heart. Many of the larger veins contain valves to prevent the back flow of blood.

THE VENOUS SYSTEM

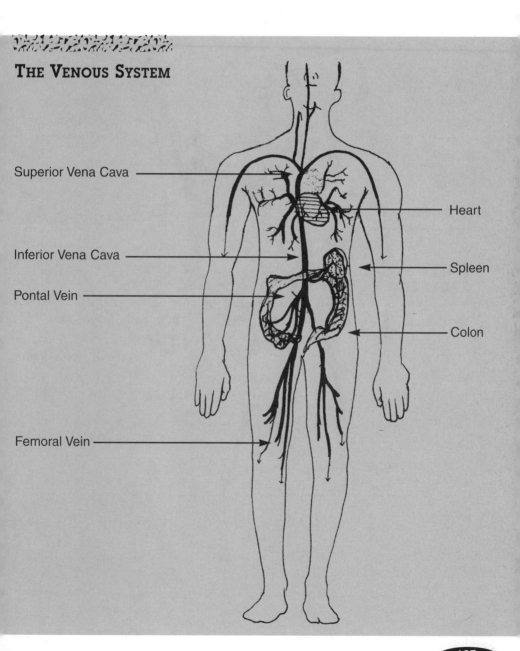

Superior Vena Cava

Inferior Vena Cava

Pontal Vein

Femoral Vein

Heart

Spleen

Colon

THE ABDOMEN

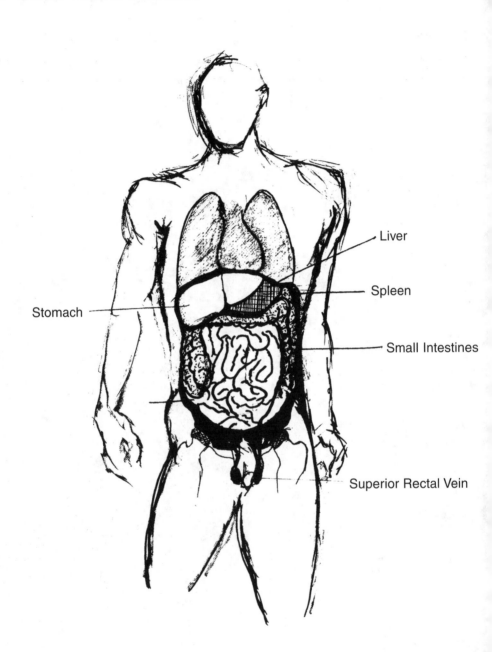

Liver

Spleen

Stomach

Small Intestines

Superior Rectal Vein

THE HEPATOPORTAL SYSTEM

The byproducts of digestion are toxic substances of a wide variety. Blood is carried from the digestive organs to the liver for cleansing through the hepatoportal system. It is a major section of the circulatory system and can carry up to one third of the body's blood.

THE HEPATOPORTAL SYSTEM

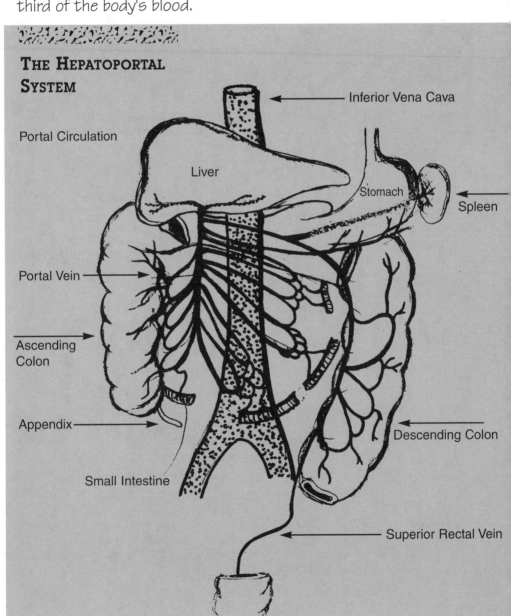

Inferior Vena Cava

Portal Circulation

Liver

Stomach

Spleen

Portal Vein

Ascending Colon

Descending Colon

Appendix

Small Intestine

Superior Rectal Vein

THE DIGESTIVE SYSTEM

Our food intake is reduced to molecules that are used by various tissues of the body through digestion. An ordinary bite of food takes about twelve hours to travel from the mouth to the end of the large intestines.

The teeth begin by cutting and grinding food. Enzymes released by salivary glands in the mouth begin the chemical breakdown of food. The food travels down the esophagus and enters the stomach. The stomach has walls of smooth muscle that expand as it is filled. Acids secreted by the stomach walls continue to break down the food. The food then travels through the small intestines, a convoluted tube divided into segments. Chemicals from the pancreas, liver, and gallbladder continue the process of food breakdown in the first segment of the small intestines. Nutrients are removed from the food in the small intestines through its walls and distributed to the bloodstream. Food particles that are still unused travel into the large intestines where the last nutrients and fluids are removed before the waste products are excreted. The length of time involved in the entire process varies widely, but is on the average a twelve-hour journey.

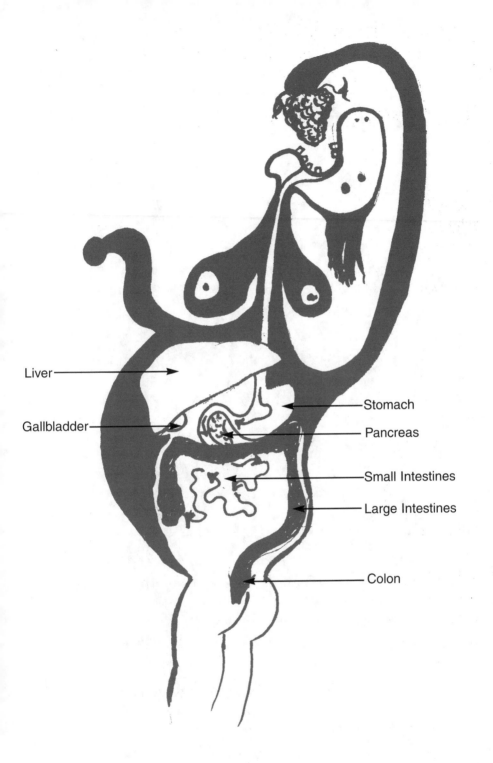

Liver

Gallbladder

Stomach

Pancreas

Small Intestines

Large Intestines

Colon

HOMEOSTASIS

The body composition operates as a dynamic steady state. With continuous shifting components, body composition is internally regulated. This composition is not absolutely constant, but held within limits the body can tolerate without an interruption in bodily function. This internal stabilization was first recognized by the French physiologist, Claude Bernard (1813-1878) in the 19th century and developed by a Harvard professor of physiology, Walter B. Cannon (1871-1945), who called this principle *homeostasis*.

Homeostatic activity is maintained by the coordinated activity of many body systems (the nervous and endocrine systems and the organs like kidneys, lungs, the alimentary canal and skin) that are places of exchange with the external environment. Oxygen, food, minerals and other substances that make up our body fluids enter through these organs. It is here that water is exchanged and wastes are eliminated.

THE PELVIS AND PERINEUM

THE UROGENITAL DIAPHRAGM

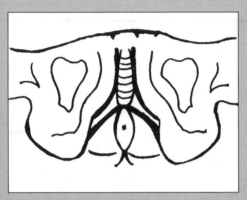
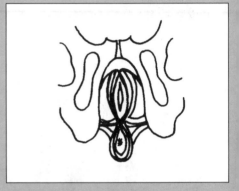

THE PELVIC DIAPHRAGM

Male

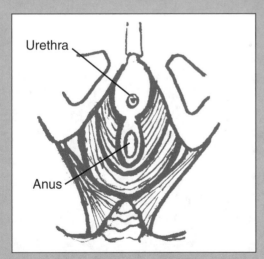

Urethra

Anus

Female

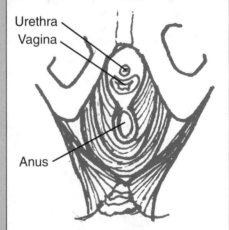

Urethra
Vagina

Anus

THE URINARY SYSTEM

As tissues convert food to energy, waste products are formed. The lungs dispose of carbon dioxide, the liver detoxifies the blood, and the urinary system eliminates most other waste products and controls fluid balance within the body.

THE LIVER

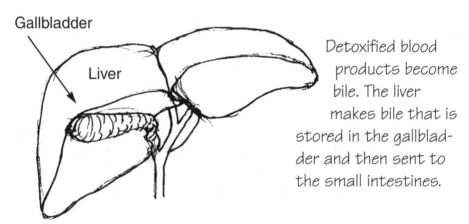

Gallbladder

Liver

Detoxified blood products become bile. The liver makes bile that is stored in the gallbladder and then sent to the small intestines.

THE URINARY TRACT

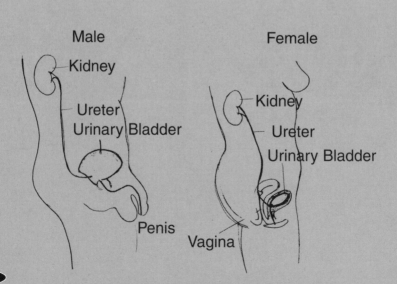

Male

Kidney

Ureter
Urinary Bladder

Penis

Female

Kidney

Ureter

Urinary Bladder

Vagina

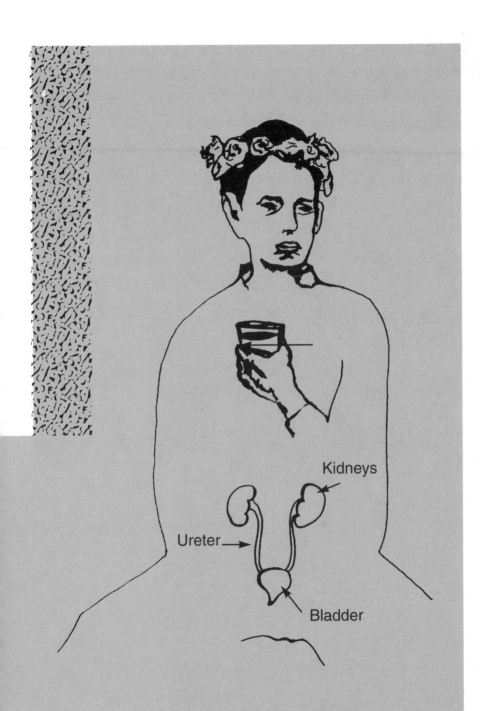

Kidneys

Ureter →

Bladder

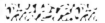

THE KIDNEY

The kidneys are the major filtering organs of the urinary system. The bladder and collecting sac are also part of the system that includes ureters which connect the kidneys to the bladder. A kidney receives unfiltered blood and extracts the impurities, excreting the liquid called urine.

NEPHRONS

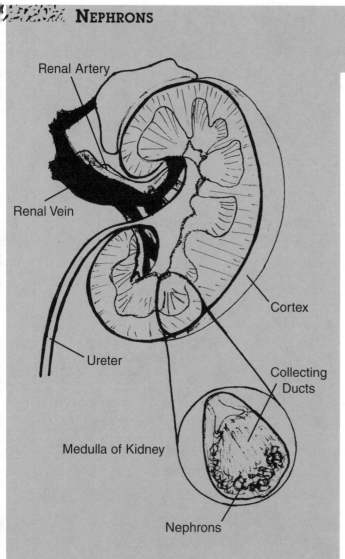

Renal Artery

Renal Vein

Cortex

Ureter

Collecting Ducts

Medulla of Kidney

Nephrons

Many structures known as nephrons filter the blood within the kidneys. Blood, carrying impurities, enter the nephrons through a small bundle of blood vessels, called glomerulus. Large particles (like blood cells) stay within the glomerulus and continue through the blood stream. The smaller unwanted particles (salts and other fluids) pass through the walls and are excreted in the urine.

THE NERVOUS SYSTEM

W e have described homeostasis as internal stabilization. It is achieved at the cell (the basic building blocks), organ, and system levels by transporting materials and energy. Homeostasis is also controlled by feedback mechanisms. Reactions to stimuli and maintenance of homeostasis require communication between cells and organs in an organism. The more complex the animal's organization, the more complex its communication between cells and organs become. The communication between the different parts of an animal are neural and hormonal. When immediate responses are required, communication occurs by neural mechanisms. These involve the movement of electrochemical changes in cell membranes. This function is carried out by the complex nervous system, which is composed of nervous tissue and small amounts of connective tissue. It is divided into the central nervous system (CNS) and peripheral nervous system (PNS).

Nervous tissue neurons (nerve cells) transmit neural pulses. Neurological cells (support cells) provide assistance and nourishment to the neurons. The nervous system reacts to changes in the external and internal environment. It controls and integrates activities of various parts of the body (i.e., heart and lungs)

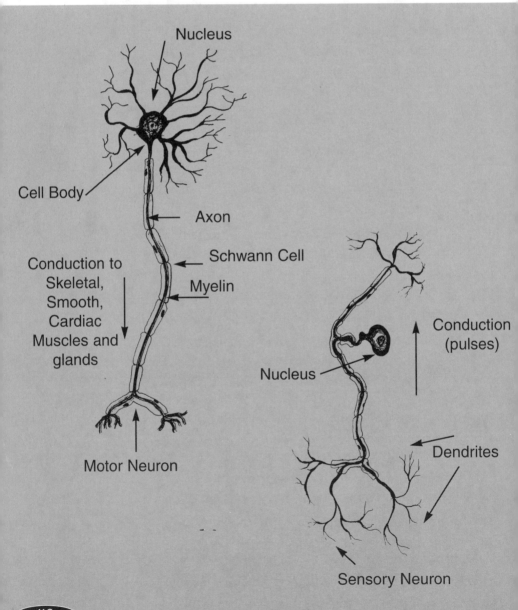

Nucleus

Cell Body

Axon

Schwann Cell

Myelin

Conduction to Skeletal, Smooth, Cardiac Muscles and glands

Motor Neuron

Nucleus

Conduction (pulses)

Dendrites

Sensory Neuron

THE CENTRAL NERVOUS SYSTEM (CNS)

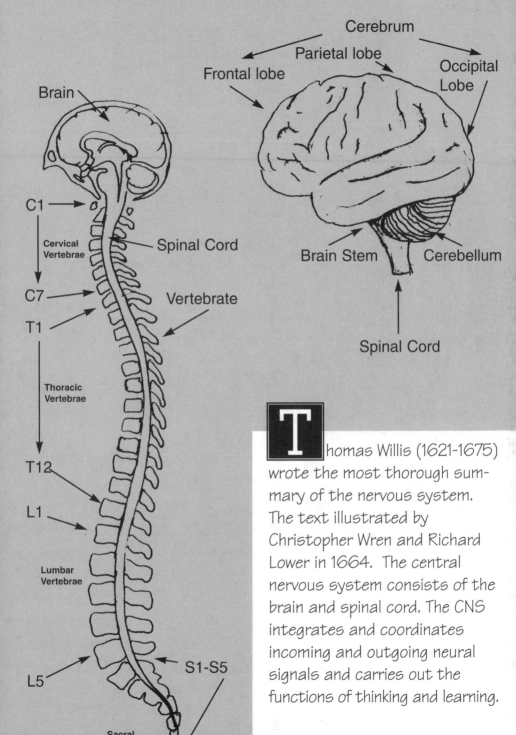

Cerebrum

Parietal lobe

Frontal lobe

Occipital Lobe

Brain

C1

Cervical Vertebrae

Spinal Cord

C7

T1

Vertebrate

Thoracic Vertebrae

Brain Stem

Cerebellum

Spinal Cord

T12

L1

Lumbar Vertebrae

L5

S1-S5

Sacral Vertebrae

Thomas Willis (1621-1675) wrote the most thorough summary of the nervous system. The text illustrated by Christopher Wren and Richard Lower in 1664. The central nervous system consists of the brain and spinal cord. The CNS integrates and coordinates incoming and outgoing neural signals and carries out the functions of thinking and learning.

117

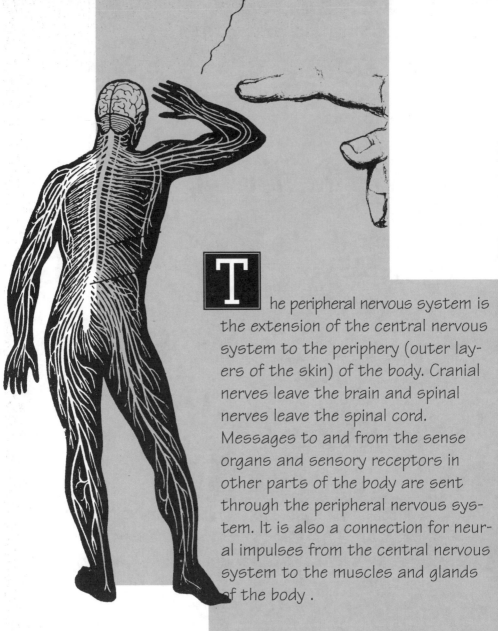

The peripheral nervous system is the extension of the central nervous system to the periphery (outer layers of the skin) of the body. Cranial nerves leave the brain and spinal nerves leave the spinal cord. Messages to and from the sense organs and sensory receptors in other parts of the body are sent through the peripheral nervous system. It is also a connection for neural impulses from the central nervous system to the muscles and glands of the body .

A bundle of nerve fibers or axons in the PNS is called a nerve. They are strong vanilla-colored cords arranged in fascicles and held together by connective tissue sheath. A network of nerves is called a nerve plexus and a collection of nerve cells is called a ganglion.

NERVE ANATOMY

Nerves have axons with a myelin sheath and neurilemmal sheath. Components of all but the smallest peripheral nerves are arranged in nerve bundles (fasciculi). Three connective tissue coverings help protect and make the delicate nerves strong. The epineurium encloses the entire nerve. It is a thick sheath of connective tissue made up of fatty tissue, blood vessels, and lymphatics. A delicate connective tissue sheath covers a bundle (fasciculus) of nerve fibers. Individual fibers are surrounded by delicate covering of connective tissue called the endoneurium.

The typical spinal nerve leaves the spinal cord by two roots: a ventral root with motor fibers and a dorsal root carrying sensory fibers from cells in the spinal cord. The exiting spinal nerve divides into a dorsal branch, supplying nerve fibers to the back, and a ventral branch, supplying nerves to the limbs and anterolateral regions of the trunk. Synapses are areas of attachment between neurons or between neurons and effector organs in the CNS and PNS.

The dorsal root, ventral root, and spinal nerves contain: (1) motor (efferent) fibers from ventral horn cells of the spinal cord; (2) sensory (afferent) fibers from spinal ganglion cells; and (3) autonomic fibers.

Autonomic Nervous System (ANS)

The autonomic nervous system is a system of nerves and ganglia that distribute impulses to heart muscle, smooth muscle and glands. It receives afferent (sensory) impulses from these parts of the body.

It is composed of a sympathetic system that stimulates activity during stress situations (i.e., heart beats and blood pressure rises). The sympathetic system is located from the thoracic vertebrate T1 to the lumbar vertebrate L2 or L3. It contains a parasympathetic system that stimulates activity that conserves and restores body resources (i.e., heart beats slowly) which is located at the cranial nerves III, VII, IX, and X and the sacral nerves in the spinal cord from S2 to S4 (sometimes S2 is absent).

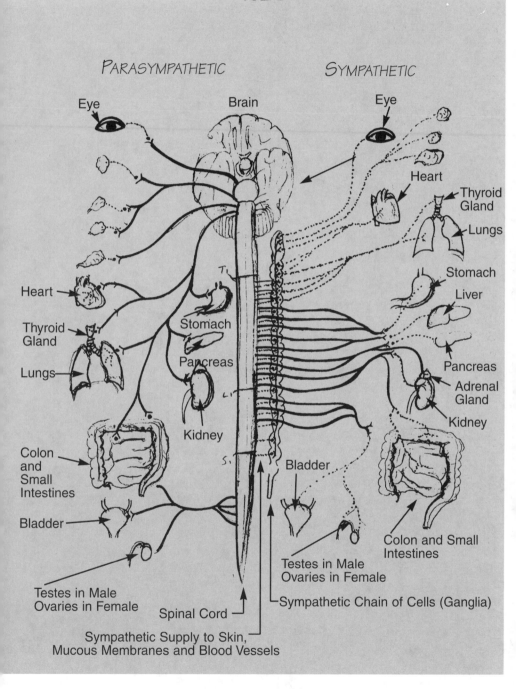

PARASYMPATHETIC

SYMPATHETIC

Eye

Brain

Eye

Heart

Thyroid
Gland

Lungs

Heart

Stomach

Thyroid
Gland

Stomach

Liver

Lungs

Pancreas

Pancreas

Adrenal
Gland

Kidney

Kidney

Colon
and
Small
Intestines

Bladder

Colon and Small
Intestines

Bladder

Testes in Male
Ovaries in Female

Testes in Male
Ovaries in Female

Sympathetic Chain of Cells (Ganglia)

Spinal Cord

Sympathetic Supply to Skin,
Mucous Membranes and Blood Vessels

THE ENDOCRINE SYSTEM

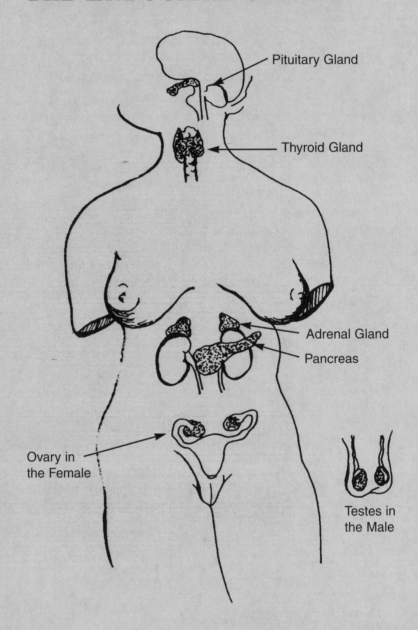

Pituitary Gland

Thyroid Gland

Adrenal Gland

Pancreas

Ovary in
the Female

Testes in
the Male

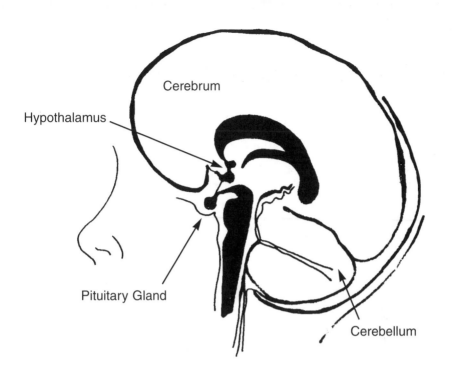

Cerebrum

Hypothalamus

Pituitary Gland

Cerebellum

T he brain signals the release of hormones that are distributed through the bloodstream. These hormones stimulate and regulate growth, sexual development, reproductive function, fluid balance, metabolism, and the response to stress. This is the endocrine system.

There is communication between the hypothalamus, the region of the brain that controls much of the hormonal system and the pituitary (a major gland lying beneath it) and the paired adrenal glands, located superior to the kidneys. The hypothalamus secretes hormones directly to the pituitary and into the blood stream. The pituitary and adrenal glands communicated via the bloodstream.

The hypothalamus is also the body's master gland that regulates the largest part of the hormonal system. Many of the hormones it produces stimulate the pituitary gland to secrete growth promoting hormones. However, one pituitary hormone decreases the female production of prolactin, preventing lactation except after birth. The hypothalamus also secretes several hormones directly into the bloodstream that do not reach the pituitary gland.

Larynx

Thyroid Gland

Trachea

Lungs

Thymus

The pituitary gland is located below and is controlled by the hypothalamus. Many of the different hormones the pituitary secretes regulate growth, others stimulate other gland function. In the pituitary, the hormone somatotropin helps control body growth before puberty, and the follicle stimulating hormone triggers the secretion of estrogens from the ovary. Estrogens are responsible for accessory sex structures: uterus, oviducts, vagina; and secondary sex characteristics: breast development, bone growth, fat deposition and hair distribution of the female.

The thyroid gland, around the windpipe, regulates the life cycle and activity levels of cells. It especially controls many of the processes used in the consumption of energy. Over-production of thyroid hormones can lead to hyperactivity. The underproduction of thyroid hormones can cause lethargy.

The adrenal glands are located just superior to the kidneys. Among their many different functions the hormones they release regulate tissue functions such as carbohydrate and electrolyte balance and contribute to sexual development. Their hormone releases are central to behavior and the fight or flight response to stress situations. These hormones also guide certain tissues in fighting disease.

Gall Bladder

Adrenal Gland

Kidneys

Pancreas

Small Intestine

The pancreas ia a long slender gland that is behind or posterior to the stomach. It is composed of two distinct types of glandular tissue. Exocrine tissues make up most of its bulk and regulate the digestion of food . Deep within the exocrine tissues are endocrine islets (islets of Langerhans) that produce many hormones. These are the glands that produce the hormone insulin, which controls much of the uptake of nutrients by the body's cells. Failure to produce adequate insulin leads to diabetes.

A hormonal substance (called pheromone) has been found to originate in one animal and effect or function as a hormone in other individuals within their group or population. These pheromones produced changes in the groups' physiological processes or behavior.

Hydra

THE NERVOUS SYSTEMS OF VERTEBRATES

When we look at the metazoans we see a gradual increase in the complexity of the nervous system. These steps of increasing complexity are probably similar to the evolutionary stages of the nervous system. A nerve net is the simplest system. It is primitive but more complex than the unorganized nervous systems of protozoans. The responses of these creatures are generalized because there is no distinction between sensory, motor or connector components. Many responses are incredibly complex for such a simple nervous system.

The beginning of the peripheral and central nervous system is thought to have been in the flatworms. They posses two anterior ganglia of nerve cells. From these two main nerve trunks run posteriorly with lateral branches extending to the other parts of the body.

Flatworm

Forebrain Midbrain Hindbrain,

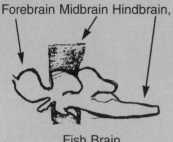

Fish Brain

Optic lobe Midbrain

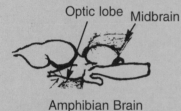

Amphibian Brain

Cerebrum

Cerebellum

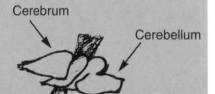

Reptile Brain

Midbrain

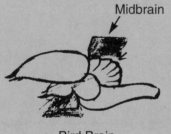

Bird Brain

Midbrain

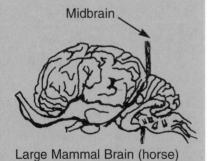

Large Mammal Brain (horse)

T he central nervous system is composed of the spinal cord and brain. Unlike the brain the structure of the vertebrate spinal cord has changed very little during its existence and evolution. The brain, however, has gone through tremendous changes. The primitive brains of fish and amphibian has become the extremely complex mammalian brain. Our brains have about 10 billion nerve cells that each connect to about 1000 others.

The primitive brain is made up of the forebrain, the midbrain, and the hindbrain (prosencephalon mesencephalon, and rhombencephalon, respectively). The forebrain and hindbrain are divided again to form the five-part brain found in all adult vertebrates. Through evolution the more advanced vertebrate has increased its powers of movement and sense of environment with the increase in the brain's complexity.

The primitive vertebrate brain was concerned with one or more special senses. And, depending on their habitats animals formed their sensory priorities. Regions of the brain were amplified for better use of a sense, like sight in the dark etc., while others were reduced.

The brain is mostly comprised of a convoluted gray matter or cortex on the outside. On the inside it is composed of white matter where myelinated bundles of nerve fibers connect the cortex, lower centers of the brain, and spinal cord, connecting one part of cortex with another. In the deep portions of the brain, clusters of nerve cells (gray matter) provide synaptic junctions between neurons of higher centers with the neurons of lower centers.

BRAIN ANATOMY & FUNCTION

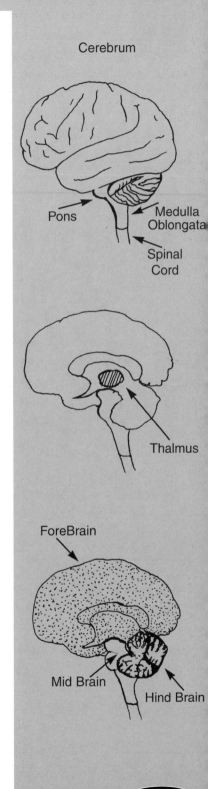

Cerebrum

Pons

Medulla Oblongata

Spinal Cord

Thalmus

ForeBrain

Mid Brain

Hind Brain

The midbrain contains the rounded optic lobes that are the centers for visual and auditory reflexes. It deals with complex behaviors in fish and amphibians and in mammals acts as a reflex center for eye muscles and a information center for sending and understanding what is heard by an animal.

In the most posterior part of the brain is the cone shaped continuation of the spinal cord called the medulla. Subconscious activities are controlled (heartbeat, breathing, swallowing) by the medulla and the more anterior midbrain. All cranial nerves, except the first (cranial nerve I, olfactory, smell), are in the brain stem. It is the most important brain area. Damage to higher centers results in debilitating loss of the senses or motor (muscle) function. Damage to the brain-stem usually results in death.

The thick bundle of nerve fibers between the medulla and the midbrain that carry impulses from one side of the cerebellum to the other is called the pons.

Above the medulla is the cerebellum. It deals with equilibrium, posture, and movement. Its evolutionary development is correlated with an animal's mode locomotion, its agility and balance. Primates have the most complex cerebellum. The human cerebellum is especially complex.

Just above the cerebellum surrounded by the massive cerebrum, the thalamus relays sensory messages coming from the spinal cord. It may be the center where signals for pain, temperature, and touch are received. There are centers in the hypothalamus that regulate body temperature, water balance, sleep, appetite and some other body functions. It also contains the cells that make the neurohormones regulate the pituitary gland.

The cerebrum can be divided into the paleocortex and the neocortex. Collectively, the paleocortex and part of the midbrain is called the limbic-midbrain system. This part of the brain has acquired functions that deal with consciousness and is concerned with sleep, memory, emotional control, and sex.

The cerebral cortex (neocortex) overshadows the paleocortex and envelops the midbrain. All integrative activities are now assigned to the neocortex.

Deep Fissure
(Longitudinal fissure)

Mammals, humans included, are said to have two brains, one primitive (all of the brain but the cerebral cortex) and one advanced (the cerebral cortex). Each controls separate functions. The deep primitive brain governs vital functions that are not under conscious control (respiration, blood pressure, etc.) This primitive brain is also the complex endocrine gland that regulates the body's endocrine system. The primitive brain is what we call the unconscious mind. The new (neo) brain, the cerebral cortex (neocortex), controls the conscious mind. It is the place of intellect and reason. Both of these brains function as one and are closely interconnected.

Posterior
(Back of Head)

Memory seems to live throughout the brain. There does not seem to be one particular region in the brain where memory can be pinpointed. The cerebral cortex contains motor and sensory areas and silent regions (association areas). Association areas are not directly connected to sense organs or muscles. They are concerned with memory, judgement and reasoning.

A deep fissure divides the cerebral cortex into two hemispheres (right and left) that are bridged by a neural mass called the corpus callosum, which allows the transfer of information and the coordination of mental activities. The right and left hemispheres of the brain are specialized for entirely different functions. The left side of the brain controls the right side of the body, as well as language develop-ment, mathematical and learning abilities. The right side controls the left side of the body as well as spatial, musical, artistic, intuitive and perceptual activities.

Corpus Callosum

THE SENSES

All organisms require the constant flow of information from the environment to regulate their survival strategies. Sense organs receive and interpret changes in the environment. The senses are excited by electrical, mechanical, chemical, and radiant forms of energy. These organs transform the messages from the environment in the form of energy into nerve impulses. Each sense organ is specifically designed to receive one kind of stimulus. Animals identify the different sensations of many kinds of stimuli in areas of the brain, where each sense organ has its own hookup. The impulses arriving at a particular sensory area of the brain are interpreted in only one way. For instance, when you put pressure on your eye, the area of the brain that receives messages for the eye tells you in terms of light sensations that your eye is being stimulated. This is the reason for the star-like patterns you see when you rub your eyes.

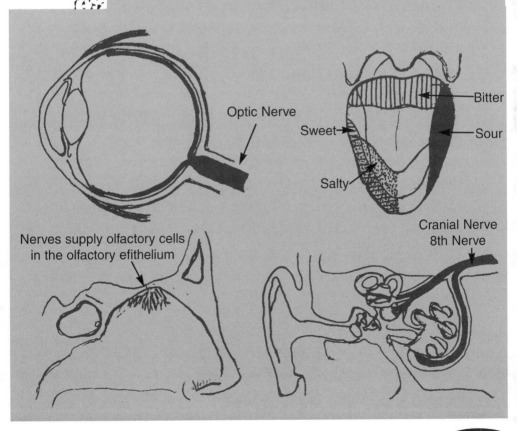

Optic Nerve

Sweet

Salty

Bitter

Sour

Nerves supply olfactory cells in the olfactory efithelium

Cranial Nerve
8th Nerve

Behavior

Environmental Responses

All life forms have the ability to be irritated. That is, they respond to the environment around them. This ability to react to the environment allows a cell or an organism to endure and reproduce. Hunting when the stomach contracts, looking for or bending toward light, avoiding light, cells eating a molecule for nutrition and avoiding poisons are all behavioral responses. When an organism cannot respond to its needs it dies.

Zoologists Karl von Frisch (Austrian-German, born 1886), Konrad Lorenz (Austrian, born 1903) and Niko Tinbergen (Dutch, born 1907) were given the Nobel Prize in physiology and medicine in 1973 for laying the foundations for the new science of ethology (character study). Ethology was once the interpretation of a person's character by observing gesture. Today, it is the scientific study of animal behavior. It is a descriptive study of the motor (movement) patterns or gestures of animals and was changed to become ethology before the middle of this century. An ethologist works to describe the behavior of an animal or groups of animals within their habitats. Many ethologists were naturalists who collected data outside.

Ethologists manipulate the variables that are provided by nature. They use animal models, play recordings of other animals, and alter animal habitats. Many of these scientists conduct experiments in laboratories where they can test their predictions under closely controlled conditions.

Today ethologists study all of the innate and learned behavior that animals use to survive and reproduce.
Most of these scientists use their laboratory observations in conjunction with observations of free-ranging animals in their natural environments. Tinbergen and Lorenz watched and catalogued the activities and vocalizations of animals during many different aspects of daily life. From feeding to mating practices, they showed that behavioral traits are measurable units like anatomical and physiological traits. They also showed that these behavioral traits have evolutionary histories. Much of behavior is determined by heredity (genes). Darwin's view of evolution also applies to behavior in that behavior evolves and is adaptive.

If a greylag goose is presented with an egg some distance from its nest, it will proceed to tuck it under its bill and roll it toward the nest. The goose continues the rolling movement until the goose reaches the nest with the egg, and rests on the nest again. If, as the goose rolls the egg, the egg escapes, the goose will continue the same pattern of using its bill as if it were rolling the egg to the nest until she is settled on the nest again. This is an example of a fixed action pattern: a motor act that is begun and remains constant throughout its performance, from start to finish. Once the goose begins the act of rolling the egg, the task must be completed whether or not the egg is present. This is an innate or instinctive skill. Fixed action patterns are completed once begun in all organisms. During the pattern of rolling the egg, the goose will make corrective movements only when the egg is actually being rolled. These corrective movements are not present when the egg rolls away or is taken. The corrective movements are orientation movements guided by environmental stimulants. Fixed action patterns usually occur with orienting movements superimposed on them.

The key environmental stimulous that triggers an animal's behavior is called a sign stimuli. Sign stimuli do not trigger fixed action patterns in all members of a species. To respond to an egg or rounded object, the goose that begins the pattern of rolling the egg must be female and in a nest. Sign stimuli are usually simple cues in the environment. These stimuli are important to the survival and reproduction of the animal. The ultrasonic cry of a bat cause night flying moths to drop to the ground and make maneuvers to avoid the bats that feed on them. The sign stimuli releases a predictable response. But it is possible that we only recognize the simplest cues. It is thought that the simplest cues reduce the chance of misinterpretation of cues.

Instinctive and Learned Behavior

The innate or instinctive behavioral pattern is released in complete form without practice or experience the first time an animal of a certain age or motivational state encounters the stimuli. When a human baby is presented with its mother's nipple, it sucks milk for the needed nutrients without practice and without experience. This act of suckling is innate or instinctive. Instinctive behaviors are driven from within the organism and are important adaptations to their survival. The changes in an animal's response is activated by internal motivations called drives (hunger drives, thirst drives, mating drives, migratory drives, etc.) This is especially true for lower animals like termites or frogs, etc., that never know their parents and must survive on their own from birth.

Advanced animals with longer lives, parental care and other social interactions are able to change previous behaviors with experience. They have the ability to learn. Habituation is learning that involves the modification or elimination of a response in the presence of reward or punishment.

Imprinting

When a gosling is strong enough to walk it can follow its mother from the nest. After following its mother for some time it will follow no other animal. Lorenz hand raised his own goslings who formed an immediate and permanent attachment. The goslings would follow and continue to follow the first large moving object they saw. Once the large object was accepted as an artificial mother they would not even follow their biological mother. This behavior is called imprinting.

Konrad Lorenz

Social Behavior

Any kind of interaction resulting from the response of one animal to another of the same species represents social behavior. Animal groupings in response to environmental signals (moths and lights) are not social situations. Social interactions depend on animal response. Living together may be beneficial to a group in terms of survival and reproduction with each species benefiting in its own particular way. Alarm calls made by an animal in a group signal predators or food. Stimulation in the form of vocalizations in some animals stimulate hormonal responses in other members of the group. Parental care and food sharing among certain animals increase the group's chance for survival as well.

The cooperation of a group depends highly on the aggressive restraints from within the group. Competition is resolved in this manner with threats or actual physical action. Animal aggression seldom results in injury or death within a species because animals have evolved many symbolic or communicative aggressive displays that are understood by others within the species.

The ecological area defended against intruders usually by the same species is their territory. When the defended space moves with the individual, it is called individual distance. Territory size is flexible and depends on the species. The resources are always defendable within a territory. Mammals are not as territorial as birds, because they can not patrol their territory often. Mammals have what we call home ranges. The home range is the total area an individual occupies during its daily activities. Usually the home range increases with body size.

Social animals are able, like humans, to communicate with one another. Though we believe it is much more restrictive than human communication, animals may communicate by sound, scent, movement and touch. They communicate with one another to hunt for food and protect home ranges or territory. Animal communication has richness and variety as exemplified in the waggle dance of the honey bee. The dance is used to communicate the direction and distance of a food source. The straight run of the waggle dance indicates direction according to the position of the sun.

Animal Cognition

D.R. Griffin (now working at Harvard University on Von Frisch's bee dance) has said that it is no more anthropomorphic to postulate mental experiences in another species than to compare its bony structure, nervous system or antibodies with our own. Studies have been done on rats, monkeys and other animals as behavioral models with the implicit assumption that principles revealed from such studies are applicable to human behavior. But the continuity of mental experiences between animals and humans is usually rejected by behaviorists.

Washoe the chimp, was taught in the 70's to use and combine gestures using the American Sign Language for the deaf. Since that time sign language studies have been extended to other chimpanzees and gorillas. Several of these animals have acquired vocabularies of several hundred reliable signs some invented by the apes themselves. In order to assess the versatility of animal communication we must first determine what sensory channel an animal is using, many of which lie undiscovered.

Chimpanzees greet each other by touching hands.

REPRODUCTIVE SYSTEM

Louis Pasteur
1822-1895

Through communicative abilities, animals are able to respond to in an appropriate way. Whether the stimuli is hormonal and picked up by smell, a physical character like color received by the eye, or a vocal mating call, animals know when it is time to reproduce. They are driven because of adaptations during their evolutionary history that cause them to choose the perfect time in their life histories to reproduce.

All life comes from life. It was once thought that life could come from inanimate matter (abiogenesis or spontaneous generation). The spontaneous generation of life was discredited by a number of scientist. Louis Pasteur showed that when a nutrient solution that carried microorganisms was boiled in a flask with an S-shaped neck, the medium was sterilized (the microbes died). Growth of the bacteria in the medium did not occur again, because the S-shaped neck prevented bacteria, living on dust particles in the air, from entering the flask. When the S-shaped neck was broken off, bacterial colonies developed quickly in the nutrient solution.

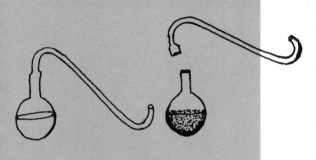

REPRODUCTION

Living organisms have the ability to make new organisms similar to themselves. Because organisms die it is necessary for them to reproduce.

Reproduction is considered by many biologists as the organisms true objective.

Reproduction practiced by most animals promotes the diversity needed in a dynamic environment. In sexual or asexual reproduction, the nutrients of the environment are converted into the offspring or sex cells that develop into offspring of similar make up. The hereditary code (DNA) is passed from the parents to the offspring.

TYPES OF REPRODUCTION

There are two fundamental types of reproduction, asexual and sexual. In asexual reproduction there is only one parent. There are no special reproductive organs or cells. Each organism that reproduces asexually can make identical copies of itself when it becomes an adult. It is simple, direct, and a quick mode of duplication. This is common among the simpler forms of life like bacteria, protozoa, etc. and is absent in the higher invertebrates.

ASEXUAL REPRODUCTION

There are four types of asexual reproduction:

1. During binary fission the body of the parent divides into two approximately equal parts. Each of these grows into an individual (Euglena).

2. Budding is an unequal division of the organism. The new individual is an outgrowth of the parent that detaches itself (i.e., Hydra).

3. In fragmentation an organism breaks into two or more parts each can then become a complete animal (i.e., starfish).

4. Multiple fission is the repeated division of the nucleus before the division of the cytoplasm. This allows the production of many daughter cells at the same time (i.e., Plasmodium vivax; a protozoa (class Sporozoea) that causes malaria in humans, is reproduced asexually in the liver cells and later in the red blood cells of the Anopheles mosquito).

Binary Fission

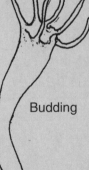

Budding

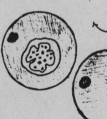

Multiple Fission

Sexual Reproduction

Most organisms in the Kingdom Animalia are sexually reproducing which involves two parents. Each contribute special sex cells (gametes) that together will develop into a new organism. The zygote formed from the gametes receives different genetic material from both parents. The combination of the genes makes a genetically unique individual with the characteristics of its species and traits with likenesses but different from its parents.

The gamete produced in the female is the ovum (egg). The ovum is produced in small numbers. It is not motile (able to move) and usually has a yolk that is the food during early development. The male produces the gamete known as spermatozoan (sperm). It is a small motile gamete and is produced in enormous numbers.

When sperm meets the egg and unites with it the egg becomes fertilized. The cell formed (the zygote) develops into the new individual.

Anatomy of the Human Reproductive System

In humans the organs that produce the germ cells or gametes are called gonads. The male gonads produce sperm and are called testes. The female gonads that produce the eggs are called ovaries. The gonads are the primary sex organs. Other accessory sex organs are present and help transfer and receive the sex cells.

THE MALE

The testes are made up of seminiferous tubules where the sperm develops. Interstitial tissue lying along the tubules produce the male sex hormone testosterone that makes sperm. The sperm move from the seminiferous tubules to the vasa efferentia to a coiled vas epididymis which is connected to the urethra by the vas deferens. They are transported through the vas deferens, where they pass the opening of the seminal vesicles, prostate glands, and Cowper's gland, where fluid is secreted by these glands to feed the sperm and lubricate the passageways. The secreted fluids also protect the sperm from the acidity of urine. They continue into the urethra in the penis and are ejaculated into the vagina during intercourse.

The external genitalia of the male is comprised of the penis, penis gland, and the scrotum that houses the testes.

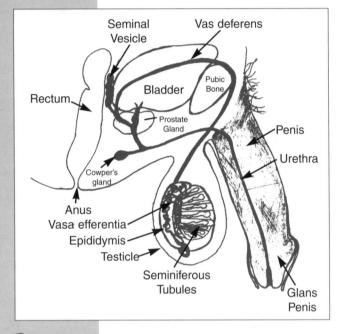

Seminal Vesicle

Vas deferens

Pubic Bone

Bladder

Rectum

Prostate Gland

Penis

Urethra

Cowper's gland

Anus

Vasa efferentia

Epididymis

Testicle

Seminiferous Tubules

Glans Penis

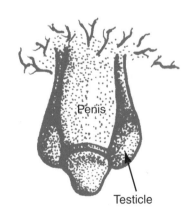

Penis

Testicle

The ovaries contain thousands of eggs (ova), and are slightly smaller than the male testes. The eggs develop in a graafian follicle that enlarges, ruptures, and releases the mature egg. When a woman is fertile approximately 13 eggs a year mature with the ovaries alternating the release of eggs. The female is only fertile for about thirty years. Only 300-400 eggs have a chance to become fertile. Oviducts or fallopian tubes, with funnel-shaped openings for receiving the eggs, carry the eggs by way of cilial propulsion. The ducts open into the upper corners of the uterus (the womb), which is specialized for housing the embryo for nine months. The uterus has thick muscular walls and many blood vessels as well as a special lining called the endometrium. The muscular tube that receives the male penis for fertilization and serves as a birth canal for the fetus is the vagina. The uterus projects down into the vagina to form the cervix.

Externally, the female genitalia or vulva is comprised of the labia majora, labia minora, and a small erectile tissue called the clitoris. The opening of the vagina is sometimes reduced in size in the virgin state by the hymen, but is often lost through physical activity during youth.

All reproductive cycles are physiologically controlled by hormones to optimize conditions for development of the young.

The female expends about 99 percent of the energy in the reproduction of new offspring but contributes only 50 percent of their hereditary material to the effort.

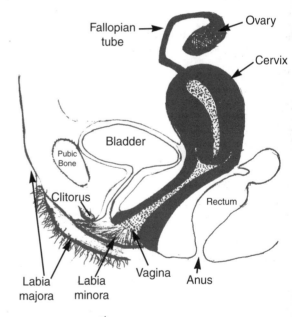

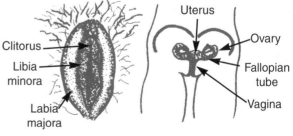

GROWTH AND HUMAN DEVELOPMENT

Growth is a fundamental characteristic of life. It is characterized by an increase in tissue or protoplasmic mass with cell division, enlargement or both. The incorporation of new materials from the environment and the availability of those materials determine growth. True developmental growth is accompanied by shaping of the embryo called morphogenesis. The specialization of the cell, tissues, and structures into the organization of the whole organism is called differentiation. During embryonic development, nuclei in different regions of the embryo provide different genetic information, resulting in the regional differentiation of cells and tissue.

DEVELOPMENT

In 1677, a Dutchman, Antony van Leeuwenhoek (1632-1723) discovered male spermatozoa with the aid of a microscope. Before this discovery it was believed by many that the embryo contained a little person that had been in the sperm or egg. People believed that the embryo just added nutrients until the fetus reached a size that could be born.

The theory of epigenesis states that an organism begins as undifferentiated material and goes through a number of stages, developing different structures until a mature embryo is formed.

Birds, reptiles and some primitive mammals leave the amniotic egg exposed. Exposed to the world it might make a good meal for other animals. Mammals have found a great way of protecting their eggs during development through the course of evolution. In marsupials (like the kangaroo), the embryo develops for some time in the uterus but does not take root in the uterine wall. The young are born immature and are sheltered and nourished in a pouch in the abdominal wall. The embryos of the more advanced mammals or placental mammals are nourished by means of a placenta.

HUMAN DEVELOPMENT

FERTILIZATION

The male and female gametes unite to form a zygote. This process of fertilization activates development. When the head of the sperm meets the egg, a fertilization cone appears, where the head of the sperm will be drawn inward. This activates a visible change that travels like a wave across the egg's surface, and a fertilization membrane is immediately elevated that prevents polyspermy (the abnormal entrance of more than one sperm). The sperm binds to the egg's surface and a layer under the plasma membrane called cortical granules break and release a material that unites and causes a build up of a new egg surface called the hyaline layer. This cortical reaction produces molecular reorganization of the cortex and removes the inhibitors that have

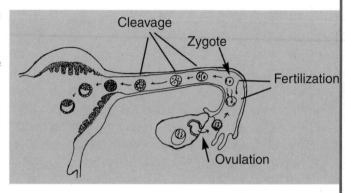

blocked the energy systems and protein synthesis in the egg that have kept it in suspended animation. The normal metabolic activity is restored the male and female pronuclei fuse and the zygote enters cleavage.

WEEKS 1 - 2

After fertilization of the human egg in the oviduct, the dividing egg travels for five days down the oviduct to the uterus. The egg is propelled by cilia and muscle peristalsis. It takes 24 hours for the first cleavage and 10 to 12 for those after the first . Cleavage produces a small ball of twenty to 30 cells (the morula) within which appears a fluid cavity (the segmentation cavity). The embryo is now a blastocyst.

The inner cell mass develops on one side of the peripheral cell (trophoblast) layer. The inner cell mass form the embryo and the surrounding trophoblast forms the placenta. At 6 days old, the blastocyst is made up of about 100 cells at which time it implants into the uterine wall (endometrium). Enzymes are produced by the trophoblasts that allow the blastocyst to sink into the endometrium . At 11 or 12 days, the blastocyst is totally buried and has eroded through the walls of capillaries and small arteriole releasing a pool of blood that bathes the embryo. Nourishment is immediately taken by direct diffusion from the blood but soon the placenta develops and takes over the function of exchange.

The outer cell trophoblast becomes the chorion and the inner cell mass becomes the fetus. The trophoblast begins to secrete human chorionic gonadotrophin (HCG). This is the hormone that is the basis for the pregnancy test. It serves to maintain the corpus luteum. The endometrium is maintained and menstruation does not occur.

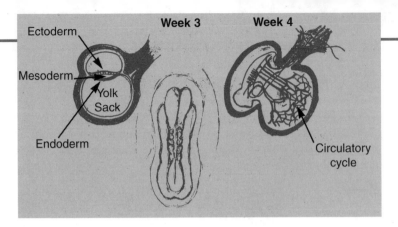

WEEK 3

The germ layers; ectoderm, endoderm, and mesoderm are laid down. The mesoderm between the ectoderm and the endoderm forms blocks of tissue along the midline called somites. Development of all structures, organs, etc., can be related to specific germ layers by this time. The ectoderm becomes the nervous system, skin, hair, and nails. The endoderm will become the inner linings of the digestive, respiratory, and urinary tracts. The mesoderm produces the muscles, skeleton, and circulatory systems. Development of the heart begins in the third week and continues, making its appearance in the fourth week.

WEEK 4

The right and left heart tubes fuse, and the heart begins to pump blood. The veins enter posteriorly, and the arteries exit anteriorly from the tubular heart. Later, the heart twists so that all major blood vessels are located anteriorly as in the adult. The umbilical cord is fully formed and connects the developing embryo to the placenta. The nervous system becomes visually evident.

WEEK 5

The limb buds appear and begin to develop into arms and legs from which form hands and feet. The fingers and toes follow soon afterward. The cartilaginous skeleton is now growing and differentiating. A skull, made of cartilage, is evident and begins to enlarge as facial features become obvious.

WEEK 6-8

The exterior is changing rapidly. As the brain develops, the head achieves its normal relationship with the body and the neck develops. The embryo's organ systems are established. At 1 1/2 inches long and the weight of an aspirin, all systems are ready, and the placenta is fully functioning. This ends the embryonic period.

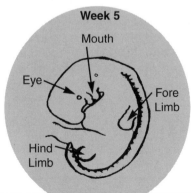

FETAL DEVELOPMENT

MONTHS 3-4

Head growth slows down and the body increases in length. The ears are now evident. The cartilage is replaced by bone, and it becomes possible to distinguish males from females. The production of the H-Y antigen causes a differentiation of the gonads into testes. If the child is to be female, the unborn has formed antibodies to this protein. In the male, the testes differentiate. They begin to produce male sex hormones and a number of androgens. One of these, testosterone, stimulates the growth of the male external genitalia. If these androgens are absent, female genitalia will form. The ovaries do not need to make estrogen because there is enough in the mother's bloodstream. At this time the testes and ovaries are located within the abdominal cavity. In the last trimester of fetal development, the testes will descend into the scrotum (scrotal sacs). The fetus is less than 6 inches and now weighs a little more than $1\frac{1}{2}$ pounds.

MONTHS 5-7

In the fifth month, the mother begins to feel the movement of the fetus. As the fetal legs grow, develop-kicks and jabs are felt. The fetus is in fetal position head down and in contact with the flexed knees. The skin is wrinkled, translucent, and pinkish in color. It is covered with a fine down called lanugo. The lanugo is coated with a white, greasy, cheese-like substance called vernix caseosa. The eyes are now fully open. At the end of this period the fetus is almost 12 inches long and weighs about 3 pounds.

The fetus does not use its lungs for gas exchange. It receives oxygen and nutrients from the mother's blood through the placenta.

MONTHS 8-9

When it is time for the birth the fetus rotates so that the head is pointing toward the cervix. A breech birth (a fetus buttocks first) sometimes calls for a caesarean section. At nine months the fetus is on the average about 21 inches and weighs around $7\frac{1}{2}$ pounds. The weight gain during this period is mostly due to an accumulation of fat under the skin. During the three stages of birth the cervix dilates, the baby is born and the afterbirth (placenta) is expelled.

12 Weeks

CELL BIOLOGY

Marcello Malpighi
(1628-1694).

T hough the magnifying glass had been around since antiquity and the eyeglass had been invented in the Middle Ages, combining the two created an instrument that was one of the most important inventions in biological science. The invention was the microscope. Antony van Leeuwenhoek, a cloth merchant who invented the microscope, and Marcello Malpighi, (1628-94) the founder of biological microscopy, made microscope lenses so well that they were made no better until the 1800s. Leeuwenhoek was the first to identify blood corpuscles. Like young children who receive a microscope Leeuwenhoek examined everything he could under his microscopes. He improved them so that they eventually had a magnification power of 270 times. He collected over four hundred microscopes by the time of his death.

Malpighi developed many techniques for preparing tissues, allowing their visibility under the microscope. He described the capillaries in the lungs that Harvey had predicted and identified the lungs as the thin-walled compartments discussed early. He also noted the small branches extending from the windpipe.

The little boxes in slices of cork and in leaves that the English scientist Robert Hooke (1635-1703) observed with an early compound microscope were some of the first cells described. He named them boxes or cells because they appeared to be compartments. All tissues and organs are composed of cells. In fact all life is composed of cells. The eggs in your refrigerator are some of the largest cells but most are invisible to the naked eye. The discovery by Matthias Schleiden in 1838 that all plants were made up of cells and Theodore Schwann's description of animal cells as being like plant cells began the new discipline of cell biology. Together these two Germans unified cell theory.

In 1931, the electron microscope was invented in Germany, and was developed in the 40's. In this microscope, high voltage drives a beam of electrons through a vacuum to make objects visible. They can detect objects only 0.001 the size of objects that can be seen with the best light microscope. Even large molecules like DNA can be seen with the electron microscope.

Antony van Leeuwenhoek
(1632-1723)

CELL STRUCTURE

In the previous sections, we began to examine structure and function at the organismic level. The invention of the microscope allows us to look at and observe the tissues of a plant or animals structure and its systems on the cellular level. It is on this level that we discover that the cell itself is the basic building block of all life.

The activities within the cell keep all things alive. All of the activities of life are caused by cell activity and cell interaction within the organism. Most cells being between .5 and 40 um in diameter have a similar shape, structure and function. Amoeba, humans, fish, and trees are all made of this single unit.

Cells are surrounded by a membrane. In plants, the cell is also surrounded by a cell wall that is secreted by the cell. All of the structures within a cell are called organelles. Some of these are enclosed by a membrane as well. The nucleus, also bound by a membrane, contains the material of heredity. Everything outside the nucleus but inside the cell's membrane is called the cytoplasm. Bacteria and blue-green algae lack organelles and a nucleus contained by a membrane. Reactions to the environment and homeostasis within all life forms require the communication between cells and organs. And the cells within a multicellular animal do communicate with and affect each other.

PROKARYOTIC AND EUKARYOTIC CELLS

Prokaryotic and eukaryotic cells have DNA, use the same genetic code and synthesize proteins. But prokaryotes lack the membrane-bound nucleus present in all eukaryotic cells.

The prokaryotic organisms are bacteria separated into the Kingdoms Archaebacteria and Monera. Blue-green algae and other bacteria are the most complex of the Prokaryotes. All other organisms are Eukaryotes. They are distributed among the four kingdoms: Protista (unicellular organisms), Fungi (true fungi), Plantae (green plants), and Animalia (multicellular animals).

EUKARYOTIC CELLS

The eukaryotic cell is enclosed by a strong, thin, membrane. It is permeable inward and outward through a membrane that regulates the flow of materials between the cell and its surroundings. Some cells, like nerve cells, have a plasma membrane that conducts communication between cells. In other cells, the plasma membrane has hundreds of small finger-like projections called microvilli. These projections increase the surface area of the cell.

The largest, most noticeable organelle within the cell is the nucleus. Unlike the prokaryotic cell's nucleus, the eukaryotic nucleus is enclosed in a double-layered membrane. The nucleus of the cell contains chromatin and nucleoli. The chromatin is a complex of DNA, the genetic information in the cell.

ORGANELLES IN THE CYTOPLASM

All of the cellular entities (except ribosomes) are composed of or enclosed within membranes. But many elements within the cytoplasm are not bound by membranes.

Microfilaments give a cell its ability to contract. Made of a special protein the alignment of the microfilaments in skeletal muscle causes the appearance of striation. They are critical to certain modes of movement in cells and their inner substances.

Microtubules are tubular in shape and a little bigger larger than microfilaments. They are made up of a protein called tubulin and play a very important role in moving the chromosomes during cell division. The shape of the cell is maintained by the supportive cytoskeleton formed by microtubules.

Centrioles do not exist in most cells of higher plants but are found in almost all animal cells. They direct the orientation of the plane of cell division. They are cylindrical and are formed by three microtubules. They also replicate before cell division.

The cells that cover the surface of a structure sometimes have moving locomotory extensions called **cilia** or **flagella**. Animals like the protozoa use these propellers to move through liquid. Flagella allow the male reproductive cells to swim in most animals and some plans. Cilia are larger and are located on most cells. Both flagella and cilia have the same inner structures. Cilia sweep water parallel to the surface of the cell and flagella propel water parallel to their main axis.

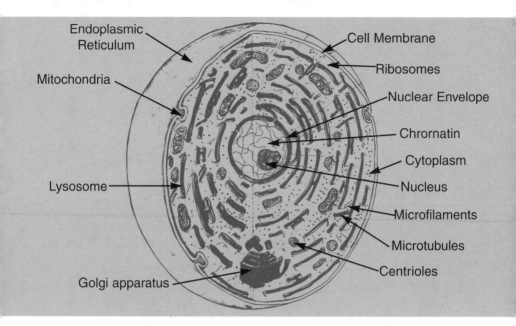

MEMBRANE BOUND ORGANELLES

The **Cisternae** is the space between membranes. The Endoplasmic Reticulum (ER) is a complex of membranes that separates the structures that make cell products from the products themselves.

The **Golgi complex** is a stack of smooth membranous cisternae that modify the storage and packaging of protein and cell products. When the products mature the ends of the cisterna that have packaged the products pinch off, becoming solitary units within the cytoplasm. Some are expelled outside of the cell. Some of these may contain digestive enzymes that stay inside the cell. These are called lysosomes. Lysosomes contain enzymes that are involved in the breakdown of foreign material (i.e., bacteria) that is stuck in the cell. These enzymes can also break down injured or diseased cells.

The **mitochondria** are some of the most strikingly obvious organelles in almost all eukaryotic cells. Their sizes, numbers, and shapes may differ greatly. They may be scattered through the cytoplasm or localized near cell surfaces and other regions, where there is constant metabolic activity (chemical processes that occur within living cells). The mitochondria have a smooth outer membrane and an inner membrane that is folded into many cristae. The mitochondria has enzymes located on the cristae that carry out the process of aerobic metabolism (ATP) that produces the energy of the cell. The mitochondria replicate themselves and have chromosomes that are almost miniatures of the chromosomes of prokaryotes. Their chromosomes contains DNA that directs the production of some of the mitochondrion proteins.

Chloroplasts are the photosynthetic organelles in plant cells.

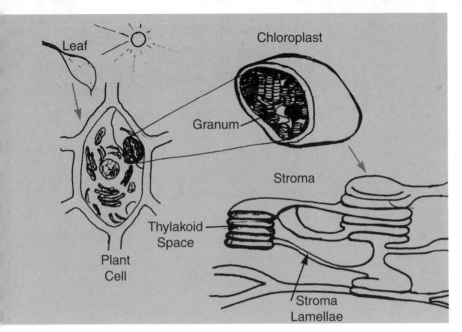

Plants and Photosynthesis

 lants, algae, and a few bacteria carry on photosynthesis. Photosynthesis is the ability to make ones own food in the presence of sunlight. All animals eat plants or eat other animals that have eaten plants.

Autotrophs possess the ability to synthesize organic molecules from inorganic raw materials and feed heterotrophs that must consume preformed organic molecules as food. Autotrophs produce organic food and heterotrophs consume it. Most life forms are ultimately dependent on the energy of the sun.

The major photosynthetic organs of a plant are its leaves, which contain many chloroplasts. These organelles, that carry on photosynthesis, are double membranous layers that surround a central space called the stroma. In the stroma enzymes incorporate (CO_2) into organic compounds. Membranes form the granum which is stacked of flattened sacs or disks called thylakoids. Grana are connected by stroma lamellae. Chlorophyll within the membrane of the grana makes them the energy-generating system of the chloroplast.

A chemical equation may be written to show that carbohydrates are the end product of photosynthesis.

$$CO_2 + H_2O + light\ energy \longrightarrow (CH_2O) + O_2$$

CELLULAR METABOLISM

A mitochondrian found within the cell

onsidered together, the chemical processes within a living cell are called cellular metabolism. During these chemical processes matter and energy are exchanged between the cell and its environment. The first law of thermodynamics tells us that energy can be stored, may change its form but cannot be lost. The second law says that all energy must at some time become totally randomized. When energy becomes more random, it can be made to perform work. The laws of thermodynamics not only apply to the physical systems, but apply to all living systems as well. The energy required for metabolism, active transport, and movement in all organisms is collected from the environment. This energy returned to the environment is less useful.

The most important energy storage molecule of all cells is produced in the mitochondria. ATP molecules are energy-rich intermediates used by life forms to power uphill reactions of cellular synthesis. Chemical reactions in cells may be either exergonic processes that proceed spontaneously with loss of free energy (energy available in a system to do work) or endergonic processes that have to be pushed uphill, because the reactants end up with more free energy than they started with.

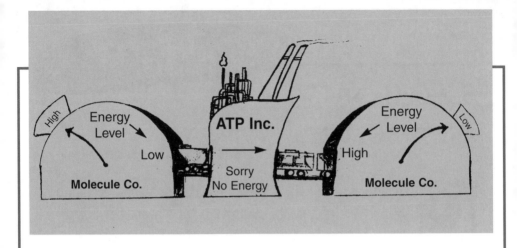

ATP (ADENOSINE TRIPHOSPHATE)

I n endergonic reactions the ATP molecule transfers chemical energy from a molecule that has a high energy content to a molecule of lower energy content. We call this a coupled reaction. ATP is the intermediate in these coupled reactions not a fuel. The ATP molecule consists of adenosine and a triphosphate group. Most of the free energy in ATP is found in the two phosphoanhydride bonds between the three phosphate groups. The bonds are called high-energy bonds because large amounts of energy in the bonds are set free when ATP is hydrolyzed to adenosine diphosphate (ADP) and inorganic phosphate. (p.103 fig. 5-3)

The production of ATP during respiration is totally dependent on the speed of its use. ATP is constructed by one set of reactions and immediately broken down by another. Living organisms do not produce and store ATP. What they store is fuel itself. The fuel is in the form of carbohydrates and lipids (fats). The production and consumption of ATP is so quick that, even in the resting cell, the ATP molecule is consumed within a minute of its formation. The average human uses 40 kilograms of ATP a day and many times more during exercise. The ATP is made as it is needed first and mostly by oxidative processes in the mitochondria. Oxygen can not be consumed unless ADP and phosphate molecules are available, and these are not available until ATP is hydrolyzed by an energy-consuming process. So metabolism is mostly self-regulating.

Almost all organisms in water or on land take in oxygen and carry on aerobic cellular respiration. Aerobic means that oxygen is required for the process. During aerobic cellular respiration, organic molecules are oxidized by the removal of hydrogen atoms. Oxidation releases the energy needed to cause ATP build up. Most of the time cellular respiration begins with glucose (sugar) but can use other molecules. When the glucose is broken down, ATP molecules are produced. During cellular respiration, glucose is oxidized to carbon dioxide (CO_2) and water (H_2O). The oxidation of glucose occurs bit by bit, not in one step.

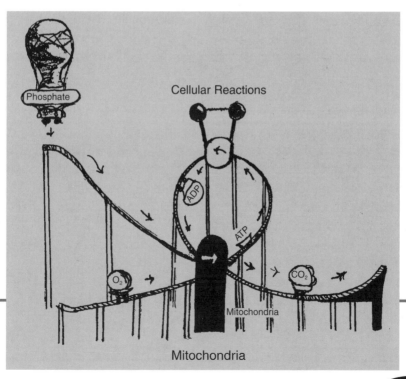

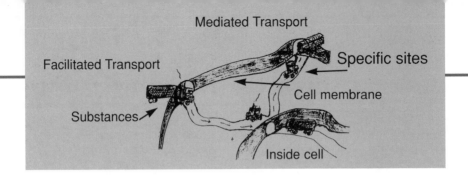

Mediated Transport

Facilitated Transport

Specific sites

Cell membrane

Substances

Inside cell

THE CELL MEMBRANE

T he cell membrane is mostly made up of protein and lipids (fats). It is the door for the entrance and exit of many substances involved in cell metabolism. Substances exit and enter the cell by free diffusion (where solutions diffuse into the cell), a mediated transport system (where a substance attaches itself to a specific site that in some way assists it across the membrane), or by endocytosis (where a substance is enclosed with a vesicle that forms on the cell and then detaches from the membrane surface to enter the cell).

I. FREE DIFFUSION

If a solute can move through a cell membrane, it will move toward the inside of the cell. The solute diffuses downhill until its concentrations are equal inside and outside the cell. Most cells are semipermeable. They are permeable to water, but not to all other solutes.

II. MEDIATED TRANSPORT

The materials needed for growth (i.e., sugar and amino acids) must enter a cell and the wastes must exit. These wastes and nutrients (molecules) are transported across the membrane barrier by specialized mechanisms built into the membrane's structure.
There are two kinds of mediated transport mechanisms, facilitated transport and active transport

IIA. FACILITATED TRANSPORT

When a molecule cannot penetrate the cell membrane, a carrier helps it diffuse through the cell membrane by facilitated transport. It helps move molecules in a downhill direction. Metabolic energy is not

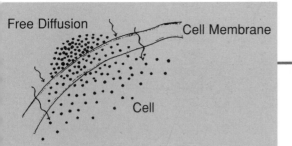

Free Diffusion

Cell Membrane

Cell

required for the carrier that helps move molecules through the cell membrane. In higher animals, facilitated transport is important for transporting glucose (blood sugar) into the body cells This is a downhill reaction. These body cells burn glucose as a principle energy source for the synthesis of ATP.

IIB. ACTIVE TRANSPORT
Active transport always involves the use or expenditure of energy (from ATP), because materials (molecules) are pumped (carried or transported) against a concentration gradient.

III. ENDOCYTOSIS
When a cell eats or ingests a solid or fluid within the cell it is called endocytosis. Pinocytosis and phagocytosis are two forms of endocytosis. They are forms of active transport because they require metabolic energy. Pinocytosis means cell drinking. When fluids are sucked through tubular channels into the cell and form tiny vesicles, some combine and form larger vacuoles. Phagocytosis means cell eating. Protozoa and lower metazoa eat this way.

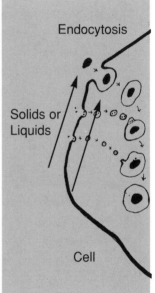

Endocytosis

Solids or Liquids

Cell

The cell membrane forms a pocket that engulfs the solid material. The membrane-enclosed vesicle then detaches from the cell surface. It moves into the cytoplasm, where the material engulfed is digested by enzymes. In this way, white blood cells (leukocytes) engulf cellular debris and uninvited microbes in the blood.

THE IMMUNE SYSTEM

Nutrients carried by fluids for the cells and tissues of the body are a perfect environment for microbes. These microbes which would cause a malfunction in the body's cells are destroyed by the immune system. The skin (integument barrier) is the first defense against microbes. When broken or somehow invaded by foreign hosts the immune system responds.

Lower invertebrates have primitive immune systems. Ameboid phagocytic cells move through their body's tissues and circulating fluids. The phagocytic cells recognize and eat (phagocytize) foreign material. They begin the repair of damaged tissue by moving to an injured area.

We have an innate nonspecific phagocytic cell system that is inherited from the invertebrates and an acquired (specific) immune response that has evolved in vertebrates. The acquired immune response provides a specific reaction to a particular infection. It is highly specific, takes time to develop and recognizes a specific foreign body and has a memory of the specific invader. It also has the ability to respond more effectively to a second attack by that foreign body.

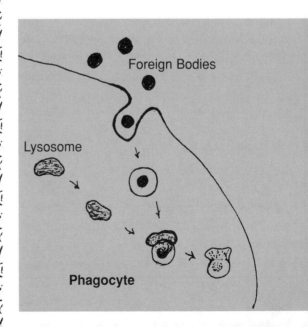

Foreign Bodies

Lysosome

Phagocyte

NONSPECIFIC DEFENSE

When an invader penetrates, the skin the cells must identify it as a foreigner. The phagocytes in all metazoa surround an invading particle, pinch it off, enclose the enemy, and destroy it with digestive enzymes supplied by lysosomes.

In metazoan invertebrates the cells performing this function are amebocytes. If the particle is too large for phagocytosis, the amebocytes may gather around it and wall it off. In humans, fixed phagocytes are found in the liver, spleen, lymph nodes, and other cleansing tissues.

Humans also have two types of mobile phagocytic cells that circulate in the blood. Leukocytes are responsible for most nonspecific phagocytic activity in the circulation. They have granules that hold the enzymes and substances to which most bacteria and fungus are toxic. Some of the substances in these granules signal the activation of cells involved in the specific immune response. Monocytes are another type of mobile phagocytic cell that move from the blood stream into the tissues in response to infection. They become activated and differentiated into active phagocytes called macrophages.

INFLAMMATION

ISLE COLLEGE RESOURCES CENTRE

Inflammation motivates the defense system against an invader. Inflammation also repairs the damage that has been caused in that area. Damaged tissues release pharmacological substances from cells in the area that increases permeability of local capillaries. Blood flow then increases in that area making it red and swollen. Neutrophils accumulate in the area and engulf the invading organisms and foreign debris. Later, macrophages predominate this area. White bloods cell may be killed in large numbers (pus formation). The invasions are quick and efficient. When the infection is destroyed, tissues repair by cell regeneration.

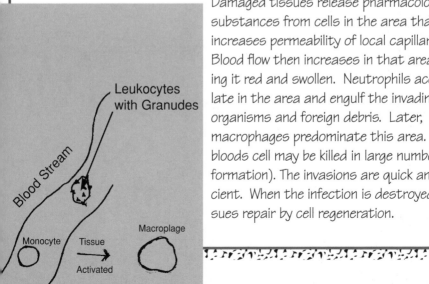

Leukocytes with Granudes

Blood Stream

Monocyte   Tissue

Activated

Macroplage

SPECIFIC IMMUNE MECHANISMS

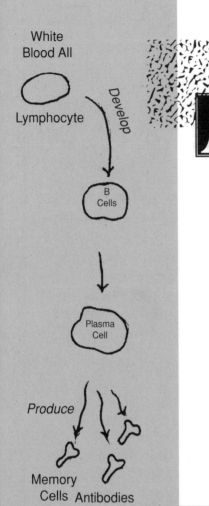

White
Blood All

Lymphocyte

Develop

B
Cells

Plasma
Cell

Produce

Memory
Cells Antibodies

Antigon

Alert
Phagocytes

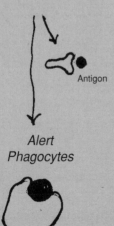

After exposure to specific foreign substances or what we call antigens, an organism produces a specific immunity. The foreign bodies are almost always proteins that stimulate an immune response. There are two immune systems involved. The humoral immunity produces antibodies that protect the organism against pathogens. The cell-mediated immunity defends the organism against parasites, viruses, and fungi. Each of these immunities is built on an unpigmented type of white blood cell called lymphocytes. Lymphocytes are made from the same tissue that makes blood in the bone marrow, but they later develop into B cells, which are responsible for humoral response and into T cells, which are responsible for cellular immune response.

HUMORAL IMMUNITY

Humoral immunity is based on antibodies. Antibodies are proteins called immunoglobulin that are produced by lymphocytes. These lymphocytes differentiate into B cells after leaving the bone marrow. The B cell matures, becoming an antibody producing cell called plasma cell. Plasma cells produce an incredible number of antibodies (about 2000 molecules/second).

160 Phagocyte

The originally sensitized lymphocytes produce plasma cells as well as long-lived memory cells. If the antigen invades again, there is what we call a secondary response. The antibody level rises quickly, 10 to 100 times the previous level. The memory cells are present so that immediate response to the re-invading antigen is possible.

The antibody level is usually undetectable when an antigen enters an animal. Only when the number of plasma cells rise can their presence be detected. The antibody level in the blood peaks and then falls after the plasma cells die and the antibodies are broken down in the system.

Antibodies can coat the surface of the invader, increasing non-specific phagocytosis; can stop the growth of antigens by causing them to stitch together; and can attach themselves to the surface of the antigen and activate the attack of a group of plasma enzymes.

The group of attacking enzymes is called a complement. These enzymes circulate in the plasma all the time in a suspended state of readiness. Activated compliments can destroy microbes, bind to antigen-antibody complexes at special sites, coat the foreign substance, and make it easier for phagocytes to eat the invader.

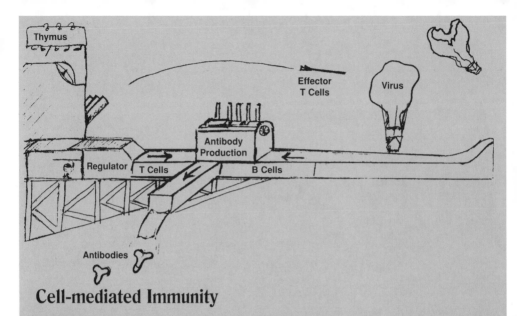

Cell-mediated Immunity

T cells are made in the thymus gland in the chest beneath the sternum. T cells look exactly like B cells, but do not make antibodies or change into antibody-producing plasma cells. Yet, they are involved in almost all immune reactions. There are many different types of T cells, which can be grouped into two categories. Effector T cells destroy target cells by direct cell-to-cell contact. The body's own cells may be targeted when they have been infected with a virus or display characteristics of a tumor-like cell. Regulator T cells interact with the B cells in the production of antibodies. Both effector T cells and regulator T cells enhance and suppress the reactions of B cells. T cells also increase the effectiveness of the body's nonspecific defense system by phagocytic cells. The immune response is regulated so that cells are available when an infection is overpowering the host. If cells are unavailable during the critical time of infection, the immune response would be too slow to stop infection.

B Cell Turnpike

Regulator and Effector T Cells Enhance and Suppress You

AIDS   ACQUIRED IMMUNE DEFICIENCY SYNDROME

All humans have the same immune system. All humans have the T cells that mediate the responses in almost all immune reactions. This is why the AIDS (HIV) virus is so deadly. It attacks the very cells that are involved in specific and non specific immune response. It is a virus. It is almost like an organism and lives only to reproduce. Through natural selection, it has evolved as a survivor. This virus has survived, because it can reproduce within the T cells of humans. The HIV virus does not discriminate; it is transferred in the fluids of the body that carry blood cells.

People do not die from the AIDS virus. They die because their immune response has failed. The immune system cannot defend the person infected from opportunistic disease. During our evolution humans have become immune to the diseases that most often attack people with the HIV virus. The diseases are usually kept under control by the immune system, almost without our knowledge. Once the immune system fails, these obscure diseases attack. These are the diseases that kill the human host. AIDS cannot be transferred by casual contact. It must involve the exchange of bodily fluids that contain a blood product infected with the AIDS virus.

The HIV virus can not be passed to another human through casual contact, food preparation, mosquitos, swimming pool water, toilet seats, etc. The AIDS virus can infect you or a sex partner through the exchange of; semen, any vaginal discharge, menstrual blood, urine, feces, and saliva with blood in it. It can also be transferred through; breast milk and other bodily fluids carrying blood cells.

GENETICS

Gregor Mendel discovered the rules of heredity. He was born the son of peasant farmers in what is now known as Czechoslovakia. His academic talents were recognized at a very early age as he was often promoted many grades at a time. At the University of Vienna he acquired his scientific education. Later he entered a monastery because his health had often failed him. The monastery provided the quiet life that Mendel needed. With his experiments Mendel discovered the rules of heredity The experiments were done at the monastery at Brunn (1856-1864), where Mendel spent much of his life. He believed that the unseen causes of inheritance were mixed when parents had offspring and mixed again when the offspring had young of their own. This was a continuous effort in nature. Because of administrative responsibilities Mendel discontinued his work on heredity at a young age. Unfortunately, the world did not know the man who died in the winter of 1884.

Gregor Mendel (1822-1884).

Martha Chase

DNA

Alfred Hershey

I n 1952, the Americans, Martha Chase and Alfred Hershey discovered that the genetic information in a cell was within the DNA molecule. In all organisms, except blue-green algae and bacteria, the hereditary material DNA is located in the nucleus. Usually found at the center of the cell, the nucleus possesses two nucleic acids that are essential to the hereditary process. They are ribonucleic acid (RNA) and deoxyribonucleic acid (DNA).

From 1951 to 1953, James D. Watson, an American, and Francis H. Crick, an Englishman, worked together with Maurice Wilkins, from London, and others and solved the puzzle of the DNA structure. In higher organisms, cells that possess nuclei DNA are found in a loose unwound state called chromatin.

James D. Watson

Francis H. Crick

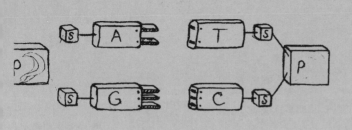

DNA COMPOSITION

Heredity is transmitted through information chemically coded (genetic code) in deoxyribonucleic acid (DNA). DNA is a two stranded molecule consisting of four chemical bases: the purines; adenine (A), and guanine (G), and the pyrimidines; cytosine (C), and thymine (T). Each is bound to another partner base. Adenine is bound to thymine. Guanine is bound to cytosine. A purine must bond with a pyrimidine.

Each base is attached to a phosphate group and a sugar molecule (deoxyribose). Together the base, phosphate and sugar are called a nucleotide. DNA is composed of a double string of nucleotides twisted into a helix-shaped spiral. Since it is two-stranded, it is called a double helix. The double helix is passed from one generation to the next by DNA replication. When a cell divides the genetic material is first replicated, making two versions of the DNA, because only one version is transmitted to each daughter cell.

DNA Replication

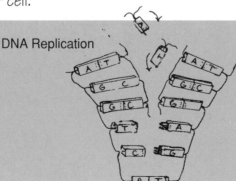

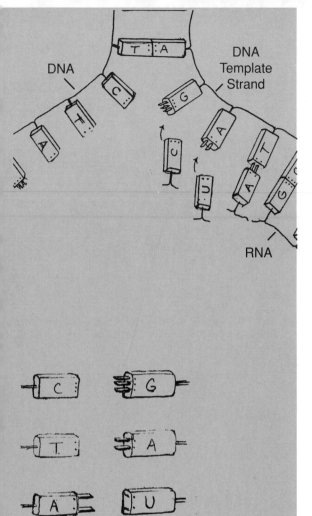

DNA

DNA Template Strand

RNA

DNA   RNA

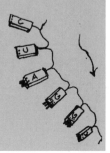

DNA
TRANSCRIPTION

DNA contains the information for all the cells in the body. The process of DNA transcription is triggered by enzymes that cause the DNA strands to unzip and partly unwind. Free-floating RNA nucleotides attach to their complementary DNA bases and a short form of RNA is made using DNA as the template.

RNA is complimentary to DNA, except that uracil (a pyrimidine) replaces thymine, which RNA does not possess. When transcription is complete, the RNA strand leaves the nucleus and moves into the cytoplasm of the cell carrying the genetic information in its sequence of bases. The RNA carrying this message is called messenger RNA (mRNA). Once in the cytoplasm, the mRNA strand attaches to a string of ribosomes (these might be called protein factories).

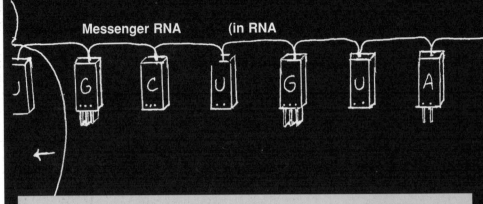

Messenger RNA   (in RNA

DNA Translation

R ibosomes move along the mRNA strand, reading or decoding the message located in the codons. Codons are three-letter symbols for the 20 common amino acids (building blocks of proteins). As the ribosome reads the mRNA, it translates the mRNA codons.

The ribosome translates by attracting transfer RNA (tRNA). Each tRNA carries a cargo fore and aft. In front is a triplet consisting of three nucleotides called an anticodon. At the rear cargo is an amino acid, and each tRNA is specific for a particular amino acid. When the ribosome reads the mRNA codon, i.e., UUU a tRNA with its complementary AAA anticodon will be attracted to the ribosome-mRNA complex, bringing along its amino acid phenylaline. The tRNA molecule plugs into the mRNA codon and then slips away leaving the amino acid behind.

The amino acid joins other amino acids left there by the same process, forming an amino acid chain, and the ribosome moves on. The completed polypeptide chain of amino acids is

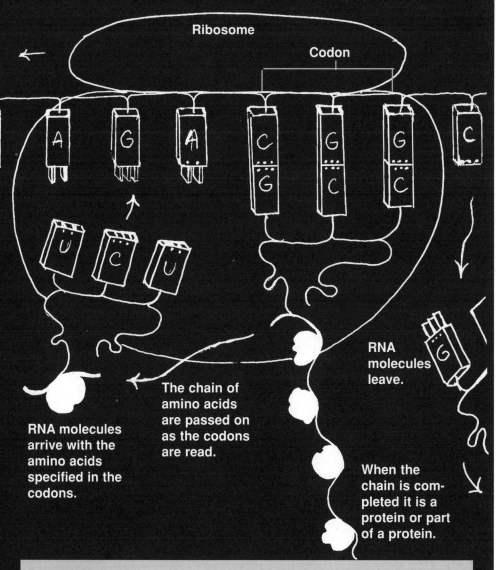

Ribosome

Codon

A G A C / G G / C G / C C

U C U

G

RNA molecules arrive with the amino acids specified in the codons.

The chain of amino acids are passed on as the codons are read.

RNA molecules leave.

When the chain is completed it is a protein or part of a protein.

a protein (or part of one) and the process of converting the DNA genetic code for the particular protein has ended.

These proteins carry out most of the metabolic functions of the cell. They also make up most of the cell structure. As we have seen, the protein synthesized depends on the information from the DNA. The result is that the protein for pigmentation, blood type, etc., is determined by the sequence of bases found in DNA.

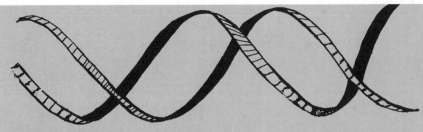

THE GENE

The gene is a section of the DNA molecule (a sequence of codons on the DNA template) that specifies the synthesis of a protein or part of a protein made up of a specific polypeptide chain of amino acids that will produce or assist in the production of one or more physical traits. It is not a unit operating in the nucleus of the cell. It is part of a large DNA molecule.

CHROMOSOMES

The long strands of DNA are found as loosely wound chromatin during the cells non-active phase. To divide and transmit the genetic material, it is condensed into compact packages called chromosomes. They are only visible when a cell is entering active division. They are two arm-like entities joined by a centromere. The number of chromosomes in different plant and animal species varies but the members of a single species posses the same number of chromosomes.

The 46 chromosomes found in the somatic cells of all humans are actually 23 pairs. Two sets of 23 different chromosomes one set from the father and one set from the mother. The first 22 pairs are called autosomes and the 23rd pair are the sex chromosomes XX or XY. A female carries the XX chromosomes and a male carries one X and one Y. Chromosomes function differently during the division of the somatic cells (mitosis) and the division of the sex cells (meiosis).

CELL DIVISION

The tissues in the body are made up of billions of somatic cells that die and are replaced by new cells. All of the cells in most multicellular organisms have originated from the division of a single cell formed from the union of the egg and sperm called the zygote.

Cell division is the basic form of growth in sexual and asexual reproduction and in the transmission of the genetic material from one cell generation to another cell generation.

Somatic cells (body cells) are formed by nuclear division, mitosis. New body cells are exactly like those they replace. Except for maturation and degeneration, the body remains essentially the same throughout life. The body cells change and take on different functions and appearances because of differential gene action.

SOMATIC CELL DIVISION: MITOSIS

Metaphase

Anaphase

Chromatia   Chromosones   Astor

Spirotes

Telophase

Two identical daughter cells w/o chromosome each.

Mitosis produces two identical daughter cells. Cells with 46 chromosomes divide into two daughter cells each containing 46 chromosomes like the mother cell. Before division, each chromosome duplicates itself, consisting of two identical strands of DNA molecules joined by a centromere. The 46 double-stranded chromosomes line up along the equator of the nucleus, and as the cell divides the centromere splits and the DNA molecules separate. As the cell continues to divide, the DNA molecules move away from each other to opposite ends of the cell. When cell division is complete, there are two daughter cells each with an identical set of 46 chromosomes (DNA molecule). Each possessing exactly the same genetic material as the mother cell. For animals that reproduce asexually, mitosis is the only sure way that the genetic information will pass from parent to offspring. In sexually reproductive animals, parents produce sex cells (gametes or germ cells) that hold only half the usual number of chromosomes, so that the offspring formed by the union of male and female sex cells will not contain double the number of parental chromosomes. We call this reductional division meiosis.

MEIOSIS

The division of a sex cell with 46 single-stranded chromosomes must produce cells called gametes that will contain one of each of the 23 chromosomes. Each gamete then contains only 23 different chromosomes. The 23 single representative chromosomes are the basic number of chromosomes and not a pair. To get to this point requires 2 cell divisions. We call these divisions Meiosis I and Meiosis II.

MEIOSIS I

During Meiosis I, each of the 46 single chromosomes (23 pairs) replicate. The cell then contains 46 double-stranded chromosomes (23 double pairs).

The members of each pair (maternal and paternal) come together and form a tetrad. The homologous chromosomes may now cross over (exchange parts). This means that sections of the original maternal and paternal chromosomes are exchanged. The material in each strand of DNA is reshuffled and will be passed on in a combination different from either parents. When the cell divides, the chromosome pairs will separate, carrying a mix of maternal and paternal hereditary material into separate cells.

The 46 double chromosomes line up randomly along the equator of the cell. After the first division, each of the daughter cells contain 23 double-stranded chromosomes.

MEIOSIS II

In Meiosis II the double stranded-molecules in each of the two cells again line up in both cells, and the strands divide. This second division produces four cells each containing 23 single stranded chromosomes.

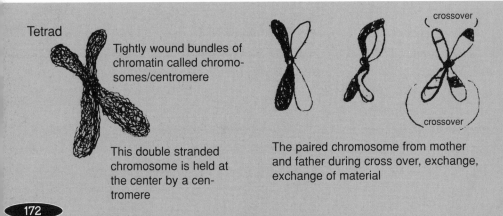

Tetrad

Tightly wound bundles of chromatin called chromosomes/centromere

This double stranded chromosome is held at the center by a centromere

crossover

crossover

The paired chromosome from mother and father during cross over, exchange, exchange of material

Here, with one chromosome, you can see how meiosis works on each chromosome within the cell.

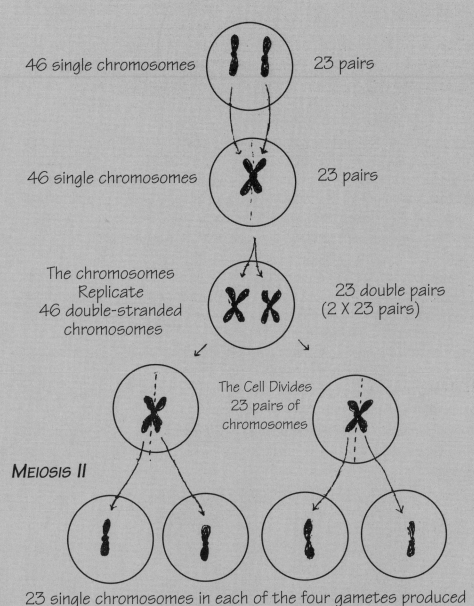

MEIOSIS I

46 single chromosomes   23 pairs

46 single chromosomes   23 pairs

The chromosomes
Replicate
46 double-stranded
chromosomes

23 double pairs
(2 X 23 pairs)

The Cell Divides
23 pairs of
chromosomes

MEIOSIS II

23 single chromosomes in each of the four gametes produced

THE GAMETES

T he result of Meiosis I and II is the division of one cell into four (gametes), each containing 23 single-stranded chromosomes. In males, these gametes are spermatids, and their formation is called spermatogenesis. When meiosis is complete, the spermatids develop tails and become mature sperm. Spermatogenesis determines the sex of the offspring. Since the male carries X and Y chromosomes, two of the four gametes carry an X and two carry a Y after Meiosis II. The chances are about even that the sperm fertilizing the ovum will carry either an X or a Y. The female ovum always carries an X because the female carries two X (XX) chromosomes and can only transmit an X chromosome. If the sperm is carrying a Y (Y from male + X from female) the offspring will be male (XY). If the sperm is carrying an X (X from male + X from female) the offspring will be female (XX).

Oogenesis (female cell division) is slightly different from spermatogenesis. The first division produces one relatively large cell and one very small cell called a polar body. In the second division the polar body divides, producing two polar bodies and the large cell divides producing one large cell and one polar body.

Meiosis produces functional gametes that begin in the ovaries of the female and testes of the male. The gametes are ready to carry the hereditary material from parents to offspring through the process of fertilization. The fertilization provides a recombination of the maternal and paternal chromosomes and restores the 23 paired chromosomes that are characteristic of our species. During fertilization the sperm also supplies the centriole needed to form a normal mitotic spindle that separates the chromosomes during division within the cell and begins the mitosis of the daughter cell.

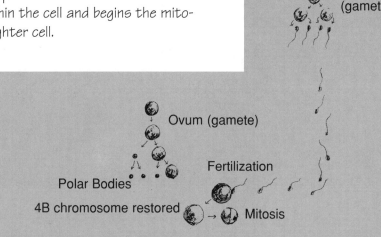

Spermatid (gametes)

Ovum (gamete)

Polar Bodies

Fertilization

4B chromosome restored

Mitosis

HEREDITY

T here are number of loci on a pair of homologous chromo At each locus on these paired homologous chromo s a gene (a segment of DNA) that affects a particular trait. In any individual, there are two genes for each trait. And there may be several loci that affect the same trait. In dealing with only one locus at a time, we are dealing with what are called simple Mendelian traits that follow simple Mendelian rules.

If, at a locus, there is a gene that contains sections of genetic material (one for smooth seeds and one for wrinkled seeds in Mendel's pea plants), then these different sections of material and forms of the gene are called alleles. Alleles are similar but slightly different sequences of DNA that control the same trait. Normal hemoglobin and sickling hemoglobin are examples of different alleles on the gene that determines what type of blood is made. For example: If male and female plant seeds are....

(The dominant trait will show up whereever its allele appears.)

Trait for seed	Male	and mate with	Female
texture	wrinkled		smooth
allele	ss		SS (dominant)

The offspring will be : Ss, Ss, Ss and Ss   all smooth seeds.

If we cross two of these daughters.... Ss   with   Ss

The offspring will be :

ss wrinkled
Ss smooth
sS smooth
SS smooth

GENETICS AND EVOLUTION

Understanding the actual mechanics of the evolutionary process comes only from understanding all the levels of organismal organization (populations, individual organisms, cells, chromosomes, DNA).

In the 1930s the "Modern Synthesis" became a new view of evolution that emphasized natural selection by gradual shifting of gene frequencies within populations. We now understand that mutations are the ultimate source of all new variation. The first source of genetic variation, of course being, the mixing of genotypes in the offspring among individuals in a population when two parents sexually reproduce.

Today, evolution is a change in allele frequency in a population (a large group of individuals) from one generation to the next. Populations change or do not change over time. In a relatively large population the alteration of one individual's genes will not significantly alter allele frequencies of the entire population. For this to happen the new allele must spread in the population.

In small popula tions, mutation in one of just a few individuals and their off-spring may alter the overall frequency quickly. This is called genetic drift and usually acts in small populations, where random factors may cause significant change in allele frequency. During human evolu-tion genetic drift may have effected our species from time to time. But most scientists believe that the lengthy evolutionary trends could only have been produced and upheld by natural selection. The way this worked in the past, and still works today, is by differential reproduction.

Individuals who carry a particular allele or combination of alleles produce more offspring. They are probably more fit for their environment. By producing more offspring than the other individuals with alternative alleles, such individu-als cause the frequency of the new allele in the population to increase slowly in proportion from generation to gener-ation. Taken over thousands of generations for millions of genes, the result is unmistakably evolutionary change. All levels of an organismal organization (molecular, cellular, individual, and populational) are interwoven in ways that can eventually produce evolutionary change.

Origin Of Life

All we have discussed briefly explains this thing we call life. But where did life come from and how did it begin? After the formation of the solar system and planets, the Earth began to cool, form a crust, condense and collect water on the surface. The earliest evidence of our planets beginnings, found in Greenland, shows that about 3,800,000,000 years ago the Earth had developed a solid crust. The pebbles and cobbles tell us that water was on the surface of the Earth to roll the rocks smooth. Certain stones even reveal the presence of free uncombined oxygen (O_2) at low levels in the atmosphere.

Life's beginnings are obscure. From 4,600,000,000 to 3,500,000,000 years ago, the most important changes from nonliving to living matter occurred. It was during this time that living cells began to populate the Earth. Some scientists believe that carbonaceous chondrite meteorites 4,600,000,000 years old may have impacted with Earth and were a source of some of the organic compounds that contributed to life's beginnings here on Earth. Dated radiometrically, these meteorites are believed to be left over from the birth of the solar system.

No one really knows how life came from non living matter. Most scientist believe it happened through a series of intermediate chemical steps. Only a few scientists think life happened all at once. Over 4,600,000,000 years ago, Venus was too close to the sun, and Mars was too far away from the sun to sustain life. Earth was in just the right place. Volcanoes gave off steam and other important gases, and the sun, lightning, and the organic meteorites produced CHON. Carbon (C), Hydrogen (H), Oxygen (O), Nitrogen (N) -> (CHON)

The air was cooled at night and heated during the day.
Cooled, heated, moistened and dried, day in and day
out. Amino acids were formed in this way. There were
right-handed and left-handed amino acids. In living
organisms today, proteins are built almost exclusively
from left-handed amino acids.

Too much water can make amino acids fall apart and
during Earth's early inorganic evolution, left-handed
amino acids proved to be the stronger of the two. The
ultraviolet light from the sun can also damage amino
acids, and as the sun's heat hit the water (H_2O), it
caused vapor to rise. When water vapor is hit by sunlight, it separates and the oxygen (O_2) and hydrogen
(H_2) become free- floating molecules. The oxygen (O_2)
became (O_3) and began to protect the Earth from the
ultraviolet light of the sun. As the amino acids grew,
the sun's burning rays caused the amino acids to fold
into spheres. Eventually, one of them was able to divide
and replicate itself. This allowed life to begin and continue. The energy-rich compounds in the water fed the
cells. They grew and multiplied. Once a source of nourishment is used up life finds other sources, and so
some of the cells found sunlight beneficial. They found
a hydrogen (H_{13}) compound they could use in the water
and released oxygen (O_{13}). These organisms were very
successful and formed mattes (collectives) in the sunlit
waters for millions of years. Their existence made
advanced cells possible.

Before cells, molecule chains of amino acid molecules began to produce simple versions of proteins and chains of nucleotides. These molecules may have been the ancestors of the genetic materials DNA and RNA, carrying information for survival to new molecules.

Like drops of oil in water, molecules tend to gather into spherical groups. Chemical reactions occurred within these early groups with smaller molecules being absorbed, rearranged and released as the waste product. The reactions were soon coordinated so that energy was released when sugar was broken apart. This energy had the ability to power other chemical reactions. New organic molecules could be built and the molecule groups grew and divided. Some grew and multiplied faster than others. These became the ancestors of single- celled animals.

So there it is, life. So well researched and so many discoveries, yet still almost inexplicable. Its presence alone is still vastly astonishing. Why does life exist? Life, clearly nothing could be more elusive. We pass each other in the street a mass of atoms, molecules, cells, and tissues wrapped and supported. That we stand or move is enough. But the human organisms we see everyday are thinking, making decisions, and changing the face of the Earth. What is this thing called life? We really don't know exactly when or how it began. Why does life, against most odds, continue? Are there other planets that could support life?

For more introductory information there are museums where you can learn much about biology. There are libraries packed with information on the subject of biology. Many subdisciplines of biology have become and are becoming separate disciplines for thousands to study. There are millions of papers and books on the subject and still hundreds of thousands of people are studying life. But it is only a scratch at of what we are made. Most colleges offer degrees in biology. Some even offer concentrations in specific disciplines such as microbiology or organismic and evolutionary biology or genetics. The field of biology, especially biomedical research, offers more opportunities in research every year. And if you don't want to think about getting a Ph.D., there are many other books you will now be able to understand.

BIBLIOGRAPHY

Abercrombie, M; Hickman, M.; Johnson, M.L.; and Thain, M.
1990 (8th ed.) *The New Penguin Dictionary of Biology*, Penguin Books, London, pp. 600.

Agur, Anne M.R.
1991 (9th ed.) (1st ed. 1943) *Grants Atlas of Anatomy*, Williams and Wilkins Pub., Baltimore pp. 650.

Allen, Thomas B., Editor
1972 *The Marvels of Animal Behavior*, Published by The National Geographic Society Washington D.C. pp. 422.

The American Museum of Natural History Central Park West, N.Y., New York

Clagett, Marshall
1955 *Greek Science in Antiquity*, Books for Libraries Press, Plainview, New York pp. 217.

Clayton, Martin
1992 *Leonardo Da Vinci: The Anatomy of Man* (Drawings from the Collection of Her Majesty Queen Elizabeth II), The Museum of Fine Arts, Houston, Bulfinch Press, Little Brown and Company, Boston pp. 141.

Diamond, M.C.; Scheibel, A.B.; and Elson, L.M.
1985 *The Human Brain Coloring Book*, Harper Perennial, Harper Collins Publishers New York pp. 301.

Eibl-Eibesfeldt, Irenaus Translation: Klinghammer, Erich
1970 *Ethology: The Biology of Behavior*, Holt Rinehart and Winston, New York

Hall, A.R.
1962 (1st ed. 1954)The Scientific Revolution 1500 - 1800: *The Formation of Modern Scientific Attitude*, Beacon Press, Boston, pp. 394.

Hickman Jr., Cleveland P.; Roberts, Larry S.; and Hickman, Frances M.
1984 (7th ed.) *Integrated Principles of Zoology* Times Mirror/Mosby College Publishing, St. Louis pp. 1065.

Lewin, Roger
1982 *Thread of Life: The Smithsonian Looks at Evolution*, Smithsonian Institution, Washington D.C., pp. 256.

Lyons, Albert S. and Petrucelli II, Joseph R.
1987 (1st ed.1978) *Medicine (An illustrated History)* Abradale Press Harry N. Abrams, Inc., Publishers, New York pp. 616.

Mader, Sylvia S.
1988 (5th ed.) *Inquiry Into Life*, Wm. C. Brown Publishers Dubuque, IA pp. 802.

Miller, Jonathan and Van Loon, Borin
1982 *Darwin for Beginners*, Writers and Readers Publishing, London, pp. 175.

Moore, Keith L.
1992 (3rd ed.) (1st ed. 1980) *Clinically Oriented Anatomy*, Williams and Wilkins Pub., Baltimore, pp. 917.

Moore, Keith L.
1977 (2nd ed.) *The Developing Human: Clinically Oriented Embryology*, W.B. Saunders Company, Philadelphia, London, Toronto pp. 411.

Moorehead, Alan
1969 *Darwin and the Beagle*, Harper and Row Publishers, New York, pp. 280.

The National Museum of Natural History The Smithsonian Institution, Washington D.C.

Nelson, Harry and Jurmain, Robert
1991 (5th ed.) *Introduction to Physical Anthropology*, West Publishing Company, pp. 640.

Rohen, Johannes W. and Yokochi, Chihiro
1988 (2nd ed.), 1983 *Color Atlas of Anatomy: A photographic Study of the Human Body*, Igaku-Shoin Ltd., New York, Tokyo pp.469.

Ronan, Colin A.
1982 *Science : Its History and Development among the World's Cultures* Facts on File Publications, N.Y. New York pp. 543.

Rosenfield, Israel; Ziff, Edward; and Van Loon, Borin
1983 *DNA for Beginners*, Writers and Readers Publishing Inc., London, New York, pp. 223.

Simpson, George Gaylord; Pittendrigh, Colin S.; and Tiffany, Lewis H.
1957 *Life: An Introduction to Biology* Harcourt, Brace and Company Inc. New York pp. 845.

INDEX